10

TESI

THESES

tesi di perfezionamento in Matematica sostenuta il 17 Marzo 2008

Luigi Manca
INRIA - Sophia Antipolis
project OPALE 2004
route des Lucioles BP 93
06902 Sophia Antipolis Cedex France

Kolmogorov Operators in Spaces of Continuous Functions and Equations for Measures

Luigi Manca

Kolmogorov Operators in Spaces of Continuous Functions and Equations for Measures

EDIZIONI
DELLA
NORMALE

ISBN: 978-88-7642-336-9

To my parents

Contents

Introduction

Let us denote by H a separable Hilbert space with norm $|\cdot|$ and inner product $\langle \cdot, \cdot \rangle$ and consider the stochastic differential equation in H

$$\begin{cases} dX(t) = \big(AX(t) + F(X(t))\big)dt + B\,dW(t), & t \geq 0 \\ \\ X(0) = x \in H, \end{cases} \tag{1}$$

where $A : D(A) \subset H \to H$ is the infinitesimal generator of a strongly continuous semigroup e^{tA}, $F : D(F) \subset H \to H$ is nonlinear, $B : H \to H$ is linear and continuous and $(W(t))_{t\geq 0}$ is a cylindrical Wiener process, defined on a stochastic basis $(\Omega, \mathcal{F}, (\mathcal{F}_t)_{t\geq 0}, \mathbb{P})$ and with values in H.

We shall assume that problem (1) has a unique solution $X(t, x)$ and we denote by P_t, $t \geq 0$, the corresponding transition semigroup, which is defined by setting

$$P_t\varphi(x) = \mathbb{E}\big[\varphi(X(t, x))\big], \quad t \geq 0, \ x \in H \tag{2}$$

where $\varphi : H \to \mathbb{R}$ is a suitable function. To fix the ideas, for $k \geq 0$ we consider the space $C_{b,k}(H)$ of all continuous mappings $\varphi : H \to \mathbb{R}$ such that

$$x \to \mathbb{R}, \quad x \mapsto \frac{\varphi(x)}{1 + |x|^k}$$

is uniformly continuous and

$$\|\varphi\|_{0,k} := \sup_{x \in H} \frac{|\varphi(x)|}{1 + |x|^k} < \infty.$$

We shall write $C_b(H) := C_{b,0}(H)$. Under suitable conditions the semigroup P_t acts on $C_{b,k}(H)$.

It is well known that the function $u(t, x) := P_t\varphi(x)$ is formally the solution of the Kolmogorov equation

$$D_t u(t, x) = K_0 u(t, x), \quad u(0, x) = \varphi(x), \tag{3}$$

where K_0 is given by

$$K_0\varphi(x) = \frac{1}{2}\text{Tr}\big[BB^*D^2\varphi(x)\big] + \langle Ax + F(x), D\varphi(x)\rangle, \quad x \in H, \quad (4)$$

and B^* is the adjoint of B.

The expression (4) is formal: it requires that φ is of class C^2, that $BB^*D^2\varphi(x)$ is a trace class operator and that $x \in D(A) \cap D(F)$. So, it is convenient to define K_0 in a suitable domain $D(K_0)$.

We start by the set of *exponential functions* $\mathcal{E}_A(H)$, which consists of the linear span of the real and imaginary part of the functions

$$H \to \mathbb{C}, \quad x \mapsto e^{i\langle x,h\rangle}, \quad h \in D(A^*),$$

where $D(A^*)$ is the domain of the adjoint operator of A.

By a simple computation we see that K_0 is well defined in $\mathcal{E}_A(H)$ and

$$K_0\varphi(x) = -|Bh|^2\varphi(x) + i\left(\langle x, A^*h\rangle + \langle F(x), h\rangle\right)\varphi(x), \quad (5)$$

for any $\varphi \in \mathcal{E}_A(H)$ and $x \in D(F)$.

Notice that if $D(F) = H$ and $|F(x)| \le c(1+|x|)$ for some $c > 0$ then $K_0\varphi \in C_{b,1}(H)$.

The semigroup P_t is not strongly continuous in $C_{b,k}(H)$ in all interesting cases. It is continuous with respect to a weaker topology, see, for instance, [8, 9, 29, 31, 40].

We shall follow the approach of π-semigroups introduced by Priola. In this approach we define the infinitesimal generator K of P_t in the space $C_{b,k}(H)$ as follows

$$\begin{cases} D(K) = \left\{\varphi \in C_{b,k}(H) : \exists g \in C_{b,k}(H), \lim\limits_{t \to 0^+} \dfrac{P_t\varphi(x) - \varphi(x)}{t} = g(x), \right. \\ \qquad\qquad \left. \forall x \in H, \ \sup\limits_{t \in (0,1)} \left\|\dfrac{P_t\varphi - \varphi}{t}\right\|_{0,k} < \infty\right\} \\ K\varphi(x) = \lim\limits_{t \to 0^+} \dfrac{P_t\varphi(x) - \varphi(x)}{t}, \varphi \in D(K), \ x \in H. \end{cases} \quad (6)$$

A problem which arises naturally is to investigate the relationships between the "abstract" operator K and the "concrete" differential operator K_0 defined in (5).

Let us briefly describe how this problem has been discussed so far.

A well-known approach consists in solving equation (3) and look for an *invariant measure* v for the semigroup P_t, that is, a probability measure on H such that

$$\int_H P_t\varphi(x)v(dx) = \int_H \varphi(x)v(dx),$$

for any $t \ge 0$, $\varphi \in C_b(H)$.

It is straightforward to check that for any $p \geq 1$, $\varphi \in C_b(H)$

$$\int_H |P_t\varphi(x)|^p \nu(dx) \leq \int_H P_t(|\varphi|^p)(x)\nu(dx) = \int_H |\varphi|^p(x)\nu(dx).$$

Hence, the semigroup P_t can be uniquely extended to a strongly continuous contraction semigroup in $L^p(H; \nu)$, for any $p \geq 1$.

Let us denote by $K_p : D(K_p) \subset L^p(H; \nu) \to L^p(H; \nu)$ its infinitesimal generator.

If the invariant measure ν enjoys suitable regularity properties then K_p is an extension of K_0 and $K_p\varphi = K_0\varphi$, for any $\varphi \in \mathcal{E}_A(H)$. The next step is to prove that $(K_0, \mathcal{E}_A(H))$ is dense in $D(K)$ endowed with the graph norm. That is, that $\mathcal{E}_A(H)$ is a *core* for $(K_p, D(K_p))$.

Many papers have been devoted to this approach, see [3, 17, 19, 21, 36] and references therein.

An other approach, based on Dirichlet forms, have been proposed to solve (3) directly in the space $L^2(H; \mu)$, see [2, 30, 35, 44] and references therein. Here μ is an infinitesimally invariant measure for K_0. Differently from the previous strategy, the solution is used to construct weak solution to (1), for instance in the sense of a martingale problem as formulated in [45].

We stress that in the described strategies, equation (3) is considered in some L^p-space. Recently, increasing attention has been devoted to study Kolmogorov operators like K_0 in spaces of continuous functions. We mention here the papers [41, 42], where the stochastic equation

$$\begin{cases} dX_t = (\Delta X(t) + F(X(t)))\, dt + \sqrt{A}\, dW(t) \\ X(0) = x \in H, \end{cases} \tag{7}$$

has been considered. Here $H := L^2(0, 1)$, $W(t)$, $t \geq 0$ is a cylindrical Wiener process on H, $A : H \to H$ is a nonnegative definite symmetric operator of trace class, Δ is the Dirichlet Laplacian on $(0, 1)$, $F : H_0^1(0, 1) \to H$ is a measurable vector field of type

$$F(x)(r) = \frac{d}{dr}(\Psi \circ x)(r) + \Phi(r, x(r)), \quad x \in H_0^1(0, 1), \ r \in (0, 1).$$

Here $H_0^1(0, 1)$ denotes the usual Sobolev space in $L^2(0, 1)$ with Dirichlet boundary conditions. The associated Kolmogorov operator is

$$L\varphi(x) = \frac{1}{2}\text{Tr}\left(AD^2\varphi(x)\right) + \langle \Delta x + F(x), D\varphi(x) \rangle,$$

where $\varphi : H \to \mathbb{R}$ is a suitable cylindrical smooth function. Roughly speaking, the authors show that L can be extended to the generator of a strongly continuous semigroup in a space of weakly continuous functions weighted by a proper Lyapunov-type function. Then, they construct a Markov process which solves equation (7) in the sense of the martingale problem.

The goal of this volume is twofold. First we want to show that K is the closure (in a suitable topology) of K_0. To get our results, we need, of course, suitable regularity properties of the coefficients and a suitable choice of $D(K_0)$. Second we want to study the following equation for measures μ_t, $t \geq 0$ on H,

$$\frac{d}{dt}\int_H \varphi(x)\mu_t(dx) = \int_H K\varphi(x)\mu_t(dx), \quad \forall\, \varphi \in D(K),\ t \in [0, T], \quad (8)$$

where μ_0 is given in advance. The precise definition will be given later.

Since the operator K is abstract, it is of interest to consider the concrete equation

$$\frac{d}{dt}\int_H \varphi(x)\mu_t(dx) = \int_H K_0\varphi(x)\mu_t(dx), \quad \forall\, \varphi \in D(K_0),\ t \in [0, T]. \quad (9)$$

For this problem uniqueness is difficult and existence is easier.

Let us give an overview on some recent developments about this problem in the finite and infinite dimensional framework.

In [5] (see also [6] for the elliptic case) a parabolic differential operator of the form

$$Lu(t, x) = \partial_t u(t, x) + \sum_{i,j=1}^d a^{ij}(t, x)\partial_{x_i}\partial_{x_j}u(t, x) + \sum_{i=1}^d b^i(t, x)\partial_{x_i}u(t, x),$$

is considered. Here $(t, x) \in (0, 1) \times \mathbb{R}^d$, $u \in C_0^\infty((0, 1) \times \mathbb{R}^d)$ and $a^{ij}, b^i : (0, 1) \times R^d \to \mathbb{R}$ are suitable locally integrable functions. The authors prove that if there exists a suitable Lyapunov-type function for the operator L, then for any probability measure ν on $\mathbb{R}^d$ there exists a family of probability measures $\{\mu_t,\ t \in (0, 1)\}$ such that

$$\int_0^1 \int_{R^d} Lu(t, x)\mu_t(dx)dt = 0$$

for any $u \in C_0^\infty((0, 1)\times\mathbb{R}^d)$ and $\lim_{t\to 0}\int_{\mathbb{R}^d}\zeta(x)\mu_t(dx) = \int_{\mathbb{R}^d}\zeta(x)\nu(dx)$, for any $\zeta \in C_0^\infty(R^d)$. Uniqueness results for this class of operators have been investigated in [4].

Equations for measures in the infinite dimensional framework have been investigated in [7]. Here it has been considered a locally convex space X and an equation for measures formally written as

$$\int_X H\varphi(t, x)\mu(dx) = 0, \qquad (10)$$

where $\varphi : X \to \mathbb{R}$ is a suitable "cylindrical" function and H is an elliptic operator of the form

$$H\varphi(t, x) = \sum_{i,j=1}^{\infty} a^{ij}(t, x)\partial_{x_i}\partial_{x_j}\varphi(x) + \sum_{i=1}^{\infty} b^i(t, x)\partial_{x_i}\varphi(x),$$

$$(t, x) \in (0, 1) \times X,$$

where $a^{ij}, b^i : (0, 1) \times R^d \to \mathbb{R}$ are suitable locally μ-integrable functions. The authors showed that under certain technical conditions, it is possible to prove existence results for the measure equation (10).

To our knowledge, there are no uniqueness results for the measure equation (9) in the infinite dimensional framework .

The main novelty of this book consists in showing existence and u-niqueness of a solution for problem (9). Differently from [7], we deal with time indipendent differential operators which act on continuous functions defined on some separable Hilbert space. We consider important cases such as Ornstein-Uhlenbeck, reaction-diffusion and Burgers operators.

Let us describe the content of this volume.

In Chapter 2 we consider an abstract situation, a general stochastically continuous Markov semigroup P_t with generator K. Under a suitable assumption we prove general existence and uniqueness results for equation (8).

In Chapter 3 we consider the case when $F = 0$. In this case P_t reduces to the Ornstein-Uhlenbeck semigroup. We prove that K is the closure of K_0 in $C_b(H)$ (this result was known, see [31]) then we solve both equation (8) and equation (9).

In Chapter 4, we consider the case when F is a bounded and Lipschitz perturbation of A. We prove that K is the closure of K_0 in $C_b(H)$. The results of Chapters 3 and 4 are contained in the paper [37].

In Chapter 5, we consider the case when F is a Lipschitz perturbation of A. We prove that K is the closure of K_0 in $C_{b,1}(H)$ (this result was known in $C_{b,2}(H)$ with more regular coefficients, see [22], [31]). We prove existence and uniqueness of a solution for problems (8), (9) when[1] $\int_H (1 + |x|)|\mu_0|_{TV}(dx) < \infty$.

---

[1] Here $|\mu_0|_{TV}$ is the total variation measure of μ_0.

In Chapter 6 we consider reaction diffusion equations of the form

$$\begin{cases} dX(t,\xi) = \big[\Delta_\xi X(t,\xi) + \lambda X(t,\xi) - p\left(X(t,\xi)\right)\big] dt \\ \qquad\qquad + B dW(t,\xi), \quad \xi \in \mathcal{O}, \\ X(t,\xi) = 0, \quad t \geq 0, \ \xi \in \partial\mathcal{O}, \\ X(0,\xi) = x(\xi), \quad \xi \in \mathcal{O}, \ x \in H, \end{cases}$$

where Δ_ξ is the Laplace operator, $B \in \mathcal{L}(H)$ and p is an increasing polynomial with leading coefficient of odd degree $d > 1$. Here $\mathcal{O} = [0,1]^n$ and $H = L^2(\mathcal{O})$. We prove that K is the closure of K_0 in the space of all continuous functions $\varphi : L^{2d}(\mathcal{O}) \to \mathbb{R}$ such that

$$\sup_{x \in L^{2d}(\mathcal{O})} \frac{|\varphi(x)|}{1 + |x|^d_{L^{2d}(\mathcal{O})}} < \infty.$$

Moreover, we prove existence and uniqueness of a solution for problems (8), (9) when

$$\int_H \big(1 + |x|^d_{L^{2d}(\mathcal{O})}\big) |\mu_0|_{TV}(dx) < \infty.$$

The results of this chapter seem to be new and are contained in the submitted paper [38].

In Chapter 7 we consider the stochastic Burgers equation in the interval $[0,1]$ with Dirichlet boundary conditions perturbed by a space-time white noise

$$\begin{cases} dX = \big(D_\xi^2 X + \tfrac{1}{2} D_\xi(X^2)\big) dt + dW, \quad \xi \in [0,1], t \geq 0, \\ X(t,0) = X(t,1) = 0 \\ X(0,\xi) = x(\xi), \ \xi \in [0,1], \end{cases}$$

where $x \in L^2(0,1)$. We prove that K is the closure of K_0 in the space of all continuous functions $\varphi : L^6(0,1) \to \mathbb{R}$ such that

$$\sup_{x \in L^6(0,1)} \frac{|\varphi(x)|}{1 + |x|^8_{L^6(0,1)} |x|^2_{L^4(0,1)}} < \infty.$$

We prove existence and uniqueness of a solution for problems (8), (9) when

$$\int_H (1 + |x|^8_{L^6(0,1)} |x|^2_{L^4(0,1)}) |\mu_0|_{TV}(dx) < \infty.$$

The results of this chapter seem to be new and they are the subject of a forthcoming paper.

ACKNOWLEDGMENT. The author wishes to express his gratitude to Prof. Giuseppe Da Prato for his helpful advice and encouragement during this work.

Chapter 1
Preliminaries

1.1. Functional spaces

Let E, E' two real Banach spaces endowed with the norms $|\cdot|_E$, $|\cdot|_{E'}$.

- We denote by $\mathcal{L}(E, E')$ the Banach algebra of all linear continuous operators $T : E \to E'$ endowed with the usual norm

$$\|T\|_{\mathcal{L}(E,E')} = \sup\{|Tx|_{E'}; \ x \in E, \ |x|_E = 1\}, \quad T \in L(E, E').$$

  If $E = E'$, we shall write $\mathcal{L}(E)$ instead of $\mathcal{L}(E, E)$. If $E' = \mathbb{R}$, the space $(E, \mathbb{R})$ is the topological dual space of E, and we shall write E^* instead of $(E, \mathbb{R})$.
- We denote by $C_b(E; E')$ the Banach space of all uniformly continuous and bounded functions $f : E \to E'$, endowed with the norm

$$\|f\|_{C_b(E,E')} = \sup_{x \in E} |f(x)|_{E'}.$$

  If $E' = \mathbb{R}$, we shall write $C_b(E)$ instead of $C_b(E; \mathbb{R})$. In some cases, we shall denote the norm of $C_b(E)$ by $\|\cdot\|_0$. However, this notation will be explicitly given when it is necessary.
- The space $C_b^1(E; E')$ consists of all $f \in C_b(E; E')$ which are Fréchet differentiable with differential $Df \in C_b(E; \mathcal{L}(E, E'))$, that is $Df : E \to \mathcal{L}(E, E')$ is uniformly continuous and bounded. The space $C_b^1(E; E')$ is a Banach space with the norm

$$\|f\|_{C_b^1(E;E')} = \|f\|_{C_b(E,E')} + \|Df\|_{C_b(E,\mathcal{L}(E,E'))}.$$

- For any $k \in \mathbb{N}$, $k > 1$, the space $C_b^k(E; E')$ consists of all $f \in C_b(E; E')$ which are k-times Fréchet differentiable with uniformly continuous and bounded differentials up to the order k.
- For any $k \in \mathbb{N}$, $k > 1$, the Banach space $C_{b,k}(E)$ consists of all functions $\varphi : E \to \mathbb{R}$ such that the function $E \to \mathbb{R}$, $x \mapsto (1 + |x|^k)^{-1}\varphi(x)$ belongs to $C_b(E)$. We set $\|\varphi\|_{0,k} := \|(1 + |\cdot|_E^k)^{-1}\varphi\|_0$.

- $\mathcal{M}(E)$ denotes the space of all finite Borel measures on E. We denote by $|\mu|_{TV}$ the total variation measure of $\mu \in \mathcal{M}(E)$.
- If $V : E \to \mathbb{R}$ is a positive measurable function, $\mathcal{M}_V(E)$ is the set of all Borel measures on E such that

$$\int_E (1 + V(x))|\mu|_{TV}(dx) < \infty.$$

If $V(x) = |x|_E^k$ for some $k > 1$ we write $\mathcal{M}_k(E)$ instead of $\mathcal{M}_V(E)$.

In most of the cases, we shall work with Hilbert spaces. Let H be a separable Hilbert space of norm $|\cdot|$ and inner product $\langle \cdot, \cdot \rangle$.

The following notations are used

- $\Sigma(H)$ is the cone in $\mathcal{L}(H)$ consisting of all symmetric operators. We set
$$L^+(H) = \{T \in \Sigma(H) : \langle Tx, x \rangle \geq 0, \quad x, y \in H\}.$$

- $L_1(H)$ is the Banach space of all trace class operators endowed with the norm
$$\|T\|_1 = \mathrm{Tr}\,\sqrt{TT^*}, \quad T \in L_1(H),$$

where Tr represents the trace. We set $L_1^+(H) = L_1(H) \cap L^+(H)$.

1.1.1. Gaussian measures

Let $m \in \mathbb{R}$ and $\lambda \in \mathbb{R}^+$. The Gaussian measure of mean m and variance λ is the measure on $\mathbb{R}$

$$N_{m,\lambda}(dx) = \begin{cases} \dfrac{1}{(2\pi\lambda)^{1/2}} e^{-\frac{(x-m)^2}{2\lambda}}\, dx, & \text{if } \lambda > 0, \\ \delta_m(dx), & \text{if } \lambda = 0, \end{cases}$$

where dx is the Lebesgue measure on $\mathbb{R}$ and δ_m is the Dirac measure at m.

Let $n > 1$. We are going to define the Gaussian measure on $\mathbb{R}^n$ of mean $a \in \mathbb{R}^n$ and covariance $Q \in L^+(\mathbb{R}^n)$. Since Q is linear, symmetric and positive, there is an orthonormal basis $\{e_1, \dots, e_n\}$ and n nonnegative numbers $\{\lambda_1, \dots, \lambda_n\}$ such that

$$Qe_i = \lambda_i e_i, \quad i \in \{1, \dots, n\}.$$

Let us consider the linear transformation $R : \mathbb{R}^n \to \mathbb{R}^n$ defined by

$$x \mapsto Rx = (\langle x, e_1 \rangle, \dots, \langle x, e_n \rangle).$$

As easily seen, R is orthonormal. Let us consider the product probability measure on $\mathbb{R}^n$

$$\mu(d\xi) := \prod_{i=1}^{n} N_{(Ra)_i, \lambda_i}(d\xi_i)$$

and define the Gaussian measure $N_{a,Q}$ by

$$\int_{\mathbb{R}^n} \varphi(x) N_{a,Q}(dx) = \int_{\mathbb{R}^n} \varphi(R^*\xi)\mu(d\xi),$$

for any integrable Borel real function $\varphi : \mathbb{R}^n \to \mathbb{R}$.

If $Q \in L^+(\mathbb{R}^n)$ is strictly positive, that is $\det Q > 0$, the Gaussian measure $N(a, Q)$ can be represented by the explicit formula

$$N_{a,Q}(dx) = (2\pi)^{-n/2}(\det Q)^{-1/2} e^{-\frac{1}{2}\langle Q^{-1}(x-a), x-a\rangle} dx, \quad x \in \mathbb{R}^n.$$

When $a = 0$, we shall write N_Q instead of $N(a, Q)$.

In order to extend the notion of Gaussian measure on a Hilbert space, we need to introduce some concepts.

Let $\mathbb{R}^\infty$ be the set of all real valued sequences.[1] $\mathbb{R}^\infty$ may be identified with the product of infinite copies of $\mathbb{R}$, that is

$$\mathbb{R}^\infty = \prod_{n \in \mathbb{N}} X_n,$$

where $X_n = \mathbb{R}$, for any $n \in \mathbb{N}$. Any element of $\mathbb{R}^\infty$ is of the form $(x_n)_{n \in \mathbb{N}}$, with $x_n \in \mathbb{R}$.

$\mathcal{B}(\mathbb{R}^\infty)$ is the σ-algebra of $\mathbb{R}^\infty$ generated by the cylindrical sets

$$\{(x_n)_{n \in \mathbb{N}} \in \mathbb{R}^\infty : x_{i_1} \in A_1, \ldots, x_{i_n} \in A_n\},$$

where $n \in \mathbb{N}$, $i_1, \ldots, i_n \in \mathbb{N}$, $A_j \in \mathcal{B}(\mathbb{R})$, $j = 1, \ldots, n$.

We identify the Hilbert space H with ℓ^2, that is the set of all sequences $(x_n)_{n \in \mathbb{N}} \in \mathbb{R}^\infty$ such that

$$\sum_{n=1}^{\infty} |x_n|^2 < \infty.$$

It is easy to see that ℓ^2 is a Borel set of $\mathbb{R}^\infty$.

Now let Q be a symmetric, nonnegative, trace class operator. Briefly, $Q \in L_1^+(H)$. We recall that $Q \in L_1^+(H)$ if and only if there exists a

---

[1] In the literature it is often denoted by $\mathbb{R}^\omega$ or $\mathbb{R}^{\mathbb{N}}$.

complete orthonormal system $\{e_k\}$ in H and a sequence of nonnegative numbers $(\lambda_k)_{k\in\mathbb{N}}$ such that

$$Qe_k = \lambda_k e_k, \quad k \in \mathbb{N}$$

and

$$\operatorname{Tr} Q = \sum_{k=1}^{\infty} \lambda_k < +\infty.$$

For any $a \in H$ and $Q \in L_1^+(H)$ we define the Gaussian probability measure $N_{a,Q}$ on $\mathbb{R}^\infty$ as a product of Gaussian measures on $\mathbb{R}$, by setting

$$N_{a,Q} = \prod_{k=1}^{\infty} N_{a_k,\lambda_k}, \quad a_k = \langle a, e_k \rangle, \ k \in \mathbb{N},$$

where N_{a_k,λ_k} is the Gaussian measure on $\mathbb{R}$ with mean a_k and variance λ_k. If $a = 0$ we shall write N_Q for brevity.

After a straightforward computation, we see that

$$\int_{\mathbb{R}^\infty} \|x\|_{\ell^2}^2 N_{a,Q}(dx) = \sum_{k=1}^{\infty}(a_k^2 + \lambda_k) = \|a\|_{\ell^2}^2 + \operatorname{Tr}Q < \infty,$$

since $a \in H$ and Q is of trace class. It follows that the measure $N_{a,Q}$ is concentrated in H, *i.e.*

$$N_{a,Q}(H) = \int_H N_{a,Q}(dx) = 1.$$

For this reason, we say that $N(a, Q)$ is a Gaussian measure on H.

Let us list some useful identities. The proof is straightforward and may be found in several texts (see, for instance, [23, 24]). We have

$$\int_H \langle x, h \rangle N_{a,Q}(dx) = \langle a, h \rangle, \qquad h \in H;$$

$$\int_H \langle x - a, h \rangle \langle x - a, k \rangle N_{a,Q}(dx) = \langle Qh, k \rangle, \qquad h, k \in H;$$

$$\int_H e^{i\langle x,h \rangle} N_{a,Q}(dx) = e^{i\langle a,h \rangle - \frac{1}{2}\langle Qh,h \rangle}, \quad h \in H.$$

1.2. The stochastic convolution

Here and in what follows we assume the following hypothesis, typical of the infinite dimensional framework (see [9, 23, 26, 13])

Hypothesis 1.2.1.
(i) $A: D(A) \subset H \to H$ is the infinitesimal generator of a strongly continuous semigroup e^{tA} of type $\mathcal{G}(M, \omega)$, i.e. there exist $M \geq 0$ and $\omega \in \mathbb{R}$ such that $\|e^{tA}\|_{\mathcal{L}(H)} \leq Me^{\omega t}, t \geq 0$;
(ii) $B \in \mathcal{L}(H)$ and for any $t > 0$ the linear operator Q_t, defined by

$$Q_t x = \int_0^t e^{sA} BB^* e^{sA^*} x \, ds, \quad x \in H, \ t \geq 0 \qquad (1.1)$$

has finite trace;
(iii) $(W(t))_{t\geq 0}$ is a cylindrical Wiener process, defined on $(\Omega, \mathcal{F}, (\mathcal{F}_t)_{t\geq 0}, \mathbb{P})$ and with values in H.

We are going to define the stochastic convolution $W_A(t)$. Formally, the Wiener process $W(t), t \geq 0$ can be written as the series

$$W(t) = \sum_{k=1}^{\infty} \beta_k(s) e_k$$

where $\{e_k, k \in \mathbb{N}\}$ is an orthonormal basis for H and $\beta_k(\cdot), k \in \mathbb{N}$ are mutually indipendent brownian motions. We formally write $W_A(t)$ as the series

$$\sum_{k=1}^{\infty} \int_0^t e^{(t-s)A} Be_k d\beta_k(s). \qquad (1.2)$$

The generic term

$$\int_0^t e^{(t-s)A} Be_k d\beta_k(s),$$

is a vector valued Wiener integral, which can be defined as

$$\int_0^t e^{(t-s)A} Be_k d\beta_k(s) = \sum_{h=1}^{\infty} \int_0^t \langle e^{(t-s)A} Be_k, e_h \rangle d\beta_k(s) \, e_h.$$

It is easy to check that

$$\left| \int_0^t e^{(t-s)A} Be_k d\beta_k(s) \right|^2 = \int_0^t |e^{(t-s)A} Be_k|^2 ds.$$

Theorem 1.2.2. *Assume that Hypothesis 1.2.1 holds. Then for any $t \geq 0$ the series in (1.2) is convergent in $L^2(\Omega, \mathcal{F}, \mathbb{P}; H)$ to a Gaussian random variable denoted $W_A(t)$ with mean 0 and covariance operator Q_t, where Q_t is defined by (1.1). In particular we have*

$$\mathbb{E}[|W_A(t)|^2] = \operatorname{Tr} Q_t.$$

Proof. See, for instance, [23]. $\qquad\qquad \square$

We study now $W_A(t)$ as a function of t. To this purpose, let us introduce the space

$$C_W([0,T]; L^2(\Omega, \mathcal{F}, \mathbb{P}; H)) := C_W([0,T]; H))$$

consisting of all continuous mappings $F \colon [0,T] \to L^2(\Omega, \mathcal{F}, \mathbb{P}; H)$ which are adapted to W, that is such that $F(s)$ is $\mathcal{F}_s$–measurable for any $s \in [0,T]$. The space $C_W([0,T]; H))$, endowed with the norm

$$\|F\|_{C_W([0,T];H))} = \left(\sup_{t\in[0,T]} \mathbb{E}\left(|F(t)|^2\right) \right)^{1/2},$$

is a Banach space. It is called the space of all *mean square continuous adapted processes* on $[0,T]$ taking values on H.

Theorem 1.2.3. *Assume that Hypothesis 1.2.1 holds. Then for any $T > 0$ we have that $W_A(\cdot) \in C_W([0,T]; H)$.*

Proof. See, for instance, [23]. $\qquad\square$

Example 1.2.4 (Heat equation in an interval). Let $H = L^2(0,\pi)$, $B = I$ and let A be given by[2]

$$\begin{cases} D(A) = H^2(0,\pi) \cap H_0^1(0,\pi), \\ Ax = D_\xi^2 x, \quad x \in D(A). \end{cases} \tag{1.3}$$

A is a self–adjoint negative operator and

$$Ae_k = -k^2 e_k, \quad k \in \mathbb{N},$$

where

$$e_k(\xi) = (2/\pi)^{1/2} \sin k\xi, \quad \xi \in [0,\pi], \quad k \in \mathbb{N}.$$

Therefore in this case Q_t is given by

$$Q_t x = \int_0^t e^{2sA} x\, ds = \frac{1}{2}(e^{2tA} - I)A^{-1}x, \quad x \in H.$$

Since

$$\operatorname{Tr} Q_t = \frac{1}{2}\sum_{k=1}^\infty \frac{1 - e^{-2tk^2}}{k^2} \leq \frac{1}{2}\sum_{k=1}^\infty \frac{1}{k^2} < +\infty,$$

we have that $Q_t \in L_1^+(H)$. Therefore Hypothesis 1.2.1 is fulfilled.

---

[2] $H^k(0,\pi), k \in \mathbb{N}$ represent Sobolev spaces and $H_0^1(0,\pi)$ is the subspace of $H^1(0,\pi)$ of all functions vanishing at 0 and π.

Example 1.2.5 (Heat equation in a square). We consider here the heat equation in the square $\mathcal{O} = [0, \pi]^N$ with $N \in \mathbb{N}$. We choose $H = L^2(\mathcal{O})$, $B = I$, and set

$$\begin{cases} D(A) = H^2(\mathcal{O}) \cap H_0^1(\mathcal{O}), \\ Ax = \Delta_\xi x, \quad x \in D(A), \end{cases}$$

where Δ_ξ represents the Laplace operator.

A is a self–adjoint negative operator in H, moreover

$$Ae_k = -|k|^2 e_k, \quad k \in \mathbb{N}^N,$$

where

$$|k|^2 = k_1^2 + \cdots + k_N^2, \quad (k_1, ..., k_N) \in \mathbb{N}^N,$$

and

$$e_k(\xi) = (2/\pi)^{N/2} \sin k_1\xi \cdots \sin k_N\xi, \quad \xi \in [0, \pi]^N, \ k \in \mathbb{N}^N.$$

In this case

$$\operatorname{Tr} Q_t = \sum_{k \in \mathbb{N}^N} \frac{1}{|k|^2} \left(1 - e^{-2t|k|^2}\right) = +\infty, \quad t > 0,$$

for any $N > 1$.

Choose now $B = (-A)^{-\alpha/2}$, $\alpha \in (0, 1)$, so that

$$Bx = \sum_{k \in \mathbb{N}^N} |k|^{-\alpha} \langle x, e_k \rangle e_k.$$

Then we have

$$\operatorname{Tr} Q_t = \sum_{k \in \mathbb{N}^N} \frac{1}{|k|^{2+2\alpha}} \left(1 - e^{-2t|k|^2}\right), \quad t > 0,$$

and so, $\operatorname{Tr} Q_t < +\infty$ provided $\alpha > N/2 - 1$.

1.2.1. Continuity in time of the stochastic convolution

We assume here that Hypothesis 1.2.1 is fulfilled. We know by Theorem 1.2.3 that $W_A(\cdot)$ is mean square continuous. We want to show that $W_A(\cdot)(\omega)$ is continuous for $\mathbb{P}$–almost all ω, that is that $W_A(\cdot)$ has continuous trajectories. For this we need the following additional assumption.

Hypothesis 1.2.6. There exists $\alpha \in (0, \frac{1}{2})$ such that

$$\int_0^1 s^{-2\alpha} \operatorname{Tr} [e^{sA} C e^{sA^*}] ds < +\infty.$$

Note that Hypothesis 1.2.6 is automatically fulfilled when C is of trace–class.

We shall use the *factorization method*, (see [12]) based on the following elementary identity

$$\int_s^t (t-\sigma)^{\alpha-1}(\sigma-s)^{-\alpha}d\sigma = \frac{\pi}{\sin\pi\alpha}, \quad 0 \le s \le \sigma \le t, \qquad (1.4)$$

where $\alpha \in (0, 1)$. Using (1.4) we can write

$$W_A(t) = \frac{\sin\pi\alpha}{\pi} \int_0^t e^{(t-\sigma)A}(t-\sigma)^{\alpha-1}Y(\sigma)d\sigma, \qquad (1.5)$$

where

$$Y(\sigma) = \int_0^\sigma e^{(\sigma-s)A}(\sigma-s)^{-\alpha}BdW(s), \quad \sigma \ge 0. \qquad (1.6)$$

To go further, we need the following analytic lemma.

Lemma 1.2.7. *Let* $T > 0$, $\alpha \in (0, 1)$, $m > 1/(2\alpha)$ *and* $f \in L^{2m}(0, T; H)$. *Set*

$$F(t) = \int_0^t e^{(t-\sigma)A}(t-\sigma)^{\alpha-1}f(\sigma)d\sigma, \quad t \in [0, T].$$

Then $F \in C([0, T]; H)$ *and there exists a constant* $C_{m,T}$ *such that*

$$|F(t)| \le C_{m,T}\|f\|_{L^{2m}(0,T;H)}, \quad t \in [0, T]. \qquad (1.7)$$

Proof. Let $M_T = \sup_{t\in[0,T]} \|e^{tA}\|$ and $t \in [0, T]$. Then by Hölder's inequality we have,

$$
\begin{aligned}
|F(t)| &\le M_T \left(\int_0^t (t-\sigma)^{(\alpha-1)\frac{2m}{2m-1}}d\sigma \right)^{\frac{2m-1}{2m}} |f|_{L^{2m}(0,T;H)} \\
&= M_T \left(\frac{2m-1}{2\alpha m-1} \right)^{\frac{2m-1}{2m}} t^{\alpha-\frac{1}{2m}} |f|_{L^{2m}(0,T;H)},
\end{aligned}
\qquad (1.8)
$$

that yields (1.7). It remains to show the continuity of F. Continuity at 0 follows from (1.8). So, it is enough to show that F is continuous at any $t_0 > 0$. For $\varepsilon < \frac{t_0}{2}$ set

$$F_\varepsilon(t) = \int_0^{t-\varepsilon} e^{(t-\sigma)A}(t-\sigma)^{\alpha-1}f(\sigma)d\sigma, \quad t \in [0, T].$$

F_ε is obviously continuous on $[\varepsilon, T]$. Moreover, using once again Hölder's inequality, we find that

$$|F(t) - F_\varepsilon(t)| \le M_T \left(\frac{2m-1}{2m\alpha-1} \right)^{\frac{2m-1}{2m}} \varepsilon^{\alpha-\frac{1}{2m}} |f|_{L^{2m}(0,T;H)}.$$

Thus $\lim_{\varepsilon \to 0} F_\varepsilon(t) = F(t)$, uniformly on $[\frac{t_0}{2}, T]$, and F is continuous at t_0 as required. $\qquad \square$

Now we are ready to prove the almost sure continuity of $W_A(\cdot)$.

Theorem 1.2.8. *Assume that Hypotheses 1.2.1 and 1.2.6 hold. Let $T > 0$ and $m \in \mathbb{N}$. Then there exists a constant $C^1_{m,T} > 0$ such that*

$$\mathbb{E}\left(\sup_{t \in [0,T]} |W_A(t)|^{2m}\right) \le C^1_{m,T}. \tag{1.9}$$

Moreover $W_A(\cdot)$ is $\mathbb{P}$–almost surely continuous on $[0, T]$.

Proof. Choose $\alpha \in (0, \frac{1}{2m})$ and let Y be defined by (1.6). Then, for all $\sigma \in (0, T]$, $Y(\sigma)$ is a Gaussian random variable N_{S_σ} where

$$S_\sigma x = \int_0^\sigma s^{-2\alpha} e^{sA} Q e^{sA^*} x \, ds, \quad x \in H.$$

Set $\mathrm{Tr}\,(S_\sigma) = C_{\alpha,\sigma}$. Then for any $m > 1$ there exists a constant $D_{m,\alpha} > 0$ such that

$$\mathbb{E}\left(|Y(\sigma)|^{2m}\right) \le D_{m,\alpha} \sigma^m, \quad \sigma \in [0, T].$$

This implies

$$\int_0^T \mathbb{E}\left(|Y(\sigma)|^{2m}\right) d\sigma \le \frac{D_{m,\alpha}}{m+1} T^{m+1},$$

so that $Y(\cdot)(\omega) \in L^{2m}(0, T; H)$ for almost all $\omega \in \Omega$. Therefore, by Lemma 1.2.7, $W_A(\cdot)(\omega) \in C([0, T]; H)$ for almost all $\omega \in \Omega$. Moreover, we have

$$\sup_{t \in [0,T]} |W_A(t)|^{2m} \le \left(\frac{C_{M,T}}{\pi}\right)^{2m} \int_0^T |Y(\sigma)|^{2m} d\sigma.$$

Now (1.9) follows taking expectation. $\qquad \square$

1.2.2. Continuity in space and time of the stochastic convolution

Here we assume that the Hilbert space H coincides with the space of functions $L^2(\mathcal{O})$ where $\mathcal{O}$ is a bounded subset of $\mathbb{R}^N$. We set

$$W_A(t)(\xi) = W_A(t, \xi), \quad t \ge 0, \ \xi \in \mathcal{O}.$$

We want to prove that, under Hypothesis 1.2.9 below, $W_A(\cdot, \cdot)(\omega) \in C([0, T] \times \mathcal{O})$ for $\mathbb{P}$–almost all $\omega \in \Omega$.

Hypothesis 1.2.9.
 (i) For any $p > 1$ the semigroup e^{tA} has a unique extension to a strongly continuous semigroup in $L^p(\mathcal{O})$ which we still denote e^{tA}.
 (ii) There exist $r \geq 2$ and, for any $\varepsilon \in [0, 1]$, $C_\varepsilon > 0$ such that

$$|e^{tA}x|_{W^{\varepsilon,p}(\mathcal{O})} \leq C_\varepsilon t^{-\frac{\varepsilon}{r}}|x|_{L^p(\mathcal{O})} \quad \text{for all } x \in L^p(\mathcal{O}). \tag{1.10}$$

(iii) A and C are diagonal with respect to the orthonormal basis $\{e_k\}$, that is there exist sequences of positive numbers $\{\beta_k\}_{k\in\mathbb{N}}$ and $\{\lambda_k\}_{k\in\mathbb{N}}$ such that

$$Ae_k = -\beta_k e_k, \quad Ce_k = \lambda_k e_k, \quad k \in \mathbb{N}.$$

Moreover, $\beta_k \uparrow +\infty$ as $k \to \infty$.
(iv) For all $k \in \mathbb{N}$, $e_k \in C(\overline{\mathcal{O}})$ and there exists $\kappa > 0$ such that

$$|e_k(\xi)| \leq \kappa, \quad k \in \mathbb{N}, \; \xi \in \overline{\mathcal{O}}. \tag{1.11}$$

 (v) There exists $\alpha \in (0, \frac{1}{2})$ such that

$$\sum_{k=1}^{\infty} \lambda_k \beta_k^{2\alpha-1} < +\infty. \tag{1.12}$$

Example 1.2.10. Assume that A is the realization of an elliptic operator of order $2m$ with Dirichlet boundary conditions in $\mathcal{O}$. Then (i) holds, (ii) holds with $r = 2m$, see e.g. [1]. (iv) does not hold in general. As easily seen, it is fulfilled when $\mathcal{O} = [0, \pi]^N$.

If $\mathcal{O} = [0, \pi]$, A is as in (1.3) and $Q = I$, then Hypothesis 1.2.9 is fulfilled with $r = 2$ and $\alpha \in (0, 1/4)$.

To prove continuity of $W_A(t, \xi)$ on (t, ξ) we need an analytic lemma.

Lemma 1.2.11. *Assume that Hypothesis* 1.2.9 *holds. Let* $T > 0$, $\alpha \in (0, 1/2)$, $m > \frac{1}{\alpha}$ *and* $f \in L^{2m}([0, T] \times \mathcal{O})$. *Set*

$$F(t) = \int_0^t e^{(t-\sigma)A}(t-\sigma)^{\alpha-1}f(\sigma)d\sigma, \quad t \in [0, T].$$

Then $F \in C([0, T] \times \mathcal{O})$ *and there exists a constant* $C_{T,m}$ *such that*

$$\sup_{t\in[0,T],\xi\in\mathcal{O}} |F(t, \xi)|^{2m} \leq C_{T,m}|f|^{2m}_{L^{2m}([0,T]\times\mathcal{O})}. \tag{1.13}$$

Proof. Set $\varepsilon = \frac{1}{2}\alpha r$. Taking into account (1.10) we have that

$$|F(t)|_{W^{\varepsilon,2m}(\mathcal{O})} \leq \int_0^t (t-\sigma)^{\alpha-1}|e^{(t-\sigma)A}f(\sigma)|_{W^{\varepsilon,2m}(\mathcal{O})}d\sigma$$

$$\leq C_\varepsilon \int_0^t (t-\sigma)^{\alpha/2-1}|f(\sigma)|_{L^{2m}(\mathcal{O})}d\sigma.$$

By using Hölder's inequality and taking into account that $\frac{m(\alpha-2)}{2m-1} > -1$, we find

$$|F(t)|^{2m}_{W^{\varepsilon,2m}(\mathcal{O})} \leq C_\varepsilon \left(\int_0^t (t-\sigma)^{\frac{m(\alpha-2)}{2m-1}} d\sigma \right)^{2m-1} |f|^{2m}_{L^{2m}([0,T]\times\mathcal{O})}.$$

Since $\varepsilon > \frac{1}{2m}$ we obtain (1.13) a consequence of Sobolev's embedding theorem. $\qquad\square$

We are now ready to prove

Theorem 1.2.12. *Assume that Hypotheses 1.2.1 and 1.2.9, hold. Then $W_A(\cdot, \cdot)$ is continuous on $[0, T] \times \mathcal{O}$, $\mathbb{P}$–almost surely. Moreover, if $m > 1/\alpha$ we have*

$$\mathbb{E}\left(\sup_{(t,\xi)\in[0,T]\times\mathcal{O}} |W_A(t,\xi)|^{2m} \right) < +\infty.$$

Proof. We write $W_A(t)$ as in (1.5), where Y is given by (1.6) with $B = \sqrt{C}$. Let us prove that $Y \in L^p([0, T] \times \mathcal{O})$, $p \geq 2$, $\mathbb{P}$–almost surely. First we notice that for all $\sigma \in [0, T]$, $\xi \in \mathcal{O}$, we have, setting $Y(\sigma)(\xi) = Y(\sigma, \xi)$,

$$Y(\sigma, \xi) = \sum_{k=1}^{\infty} \sqrt{\lambda_k} \int_0^\sigma e^{-\beta_k(\sigma-s)}(\sigma - s)^{-\alpha} e_k(\xi) d\beta_k(\sigma).$$

Thus, $Y(\sigma, \xi)$ is a real Gaussian random variable with mean 0 and covariance $\gamma(\sigma, \xi) = \gamma$ given by

$$\gamma = \sum_{k=1}^{\infty} \lambda_k \int_0^\sigma e^{-2\beta_k s} s^{-2\alpha} |e_k(\xi)|^2 ds.$$

Taking into account (1.11) and (1.12) we see that

$$\gamma \leq \sum_{k=1}^{\infty} \lambda_k \int_0^{+\infty} e^{-2\beta_k s} s^{-2\alpha} |e_k(\xi)|^2 ds$$

$$= \kappa^2 2^{2\alpha-1} \Gamma(1 - 2\alpha) \sum_{k=1}^{\infty} \lambda_k \beta_k^{2\alpha-1} < +\infty.$$

Therefore there exists $C_m > 0$ such that

$$\mathbb{E}|Y(\sigma, \xi)|^{2m} \leq C_m, \qquad m > 1.$$

It follows that

$$\mathbb{E} \int_0^T \int_{\mathcal{O}} |Y(\sigma, \xi)|^{2m} d\sigma \, d\xi \leq T C_m |\mathcal{O}|,$$

where $|\mathcal{O}|$ is the Lebesgue measure of $\mathcal{O}$. So $Y \in L^{2m}([0, T] \times \mathcal{O})$ and consequently $W_A \in C([0, T] \times \mathcal{O})$, $\mathbb{P}$–a. s. Now the conclusion follows taking expectation in (1.13), with W_A replacing F and Y replacing f. $\qquad\square$

Chapter 2
Measure valued equations for stochastically continuous Markov semigroups

2.1. Notations and preliminary results

Let E be a separable Banach space with norm $|\cdot|_E$.

We recall that $C_b(E)$ is the Banach space of all uniformly continuous and bounded functions $\varphi : E \to \mathbb{R}$ endowed with the supremum norm

$$\|\varphi\|_0 = \sup_{x \in E} |\varphi(x)|.$$

Definition 2.1.1. A sequence $(\varphi_n)_{n \in \mathbb{N}} \subset C_b(E)$ is said to be π-*convergent* to a function $\varphi \in C_b(E)$ if for any $x \in E$ we have

$$\lim_{n \to \infty} \varphi_n(x) = \varphi(x)$$

and

$$\sup_{n \in \mathbb{N}} \|\varphi_n\|_0 < \infty.$$

Similarly, the m-indexed sequence $(\varphi_{n_1,\dots,n_m})_{n_1 \in \mathbb{N}, \dots, n_m \in \mathbb{N}} \subset C_b(E)$ is said to be π-convergent to $\varphi \in C_b(E)$ if for any $i \in \{2, \dots, m\}$ there exists an $i-1$-indexed sequence $(\varphi_{n_1,\dots,n_{i-1}})_{n_1 \in \mathbb{N}, \dots, n_{i-1} \in \mathbb{N}} \subset C_b(E)$ such that

$$\lim_{n_i \to \infty} \varphi_{n_1,\dots,n_i} \overset{\pi}{=} \varphi_{n_1,\dots,n_{i-1}}$$

and

$$\lim_{n_1 \to \infty} \varphi_{n_1} \overset{\pi}{=} \varphi.$$

We shall write

$$\lim_{n_1 \to \infty} \cdots \lim_{n_m \to \infty} \varphi_{n_1,\dots,n_m} \overset{\pi}{=} \varphi$$

or $\varphi_n \overset{\pi}{\to} \varphi$ as $n \to \infty$, when the sequence has one index.

It is worth noticing that if the sequence $(\varphi_{n,m})_{n,m\in\mathbb{N}} \subset C_b(E)$ is π-convergent to $\varphi \subset C_b(E)$ we can not, in general, extract a subsequence $(\varphi_{n_k,m_k})_{k\in\mathbb{N}}$ which is still π-convergent to φ. This is the reason for which we consider multi-indexed sequences. However, in order to avoid heavy notations, we shall often write a multi-indexed sequence as a sequence with only one index.

Remark 2.1.2. As easily seen the π-convergence implies the convergence in $L^p(E;\mu)$, for any $\mu \in \mathcal{M}(E)$, $p \in [1,\infty)$.

Remark 2.1.3. The notion of π-convergence is considered also in [29], under the name of *boundedly and pointwise* convergence.

Remark 2.1.4. The topology on $C_b(E)$ induced by the π-convergence is not sequentially complete. For a survey on this fact see [31, 40].

Definition 2.1.5. For any subset $D \subset C_b(E)$ we say that φ belongs to the π-closure of D, and we denote it by $\varphi \in \overline{D}^\pi$, if there exists $m \in \mathbb{N}$ and an m-indexed sequence $\{\varphi_{n_1,\dots,n_m}\}_{n_1\in\mathbb{N},\dots,n_m\in\mathbb{N}} \subset D$ such that

$$\lim_{n_1\to\infty} \cdots \lim_{n_m\to\infty} \varphi_{n_1,\dots,n_m} \stackrel{\pi}{=} \varphi.$$

Finally, we shall say that a subset $D \subset C_b(E)$ is π-dense if $\overline{D}^\pi = C_b(E)$.

2.2. Stochastically continuous semigroups

We denote by $\mathcal{B}(E)$ the Borel σ-algebra of E.

Definition 2.2.1. A family of operators $(P_t)_{t\geq 0} \subset \mathcal{L}(C_b(E))$ is a *stochastically continuous Markov semigroup* if there exists a family $\{\pi_t(x,\cdot),\ t \geq 0,\ x \in E\}$ of probability Borel measures on E such that

- the map $\mathbb{R}^+ \times E \to [0,1]$, $(t,x) \mapsto \pi_t(x,\Gamma)$ is Borel, for any Borel set $\Gamma \in \mathcal{B}(E)$;
- $P_t\varphi(x) = \int_E \varphi(y)\pi_t(x,dy)$, for any $t \geq 0$, $\varphi \in C_b(E)$, $x \in E$;
- for any $\varphi \in C_b(E)$, $x \in E$, the map $\mathbb{R}^+ \to \mathbb{R}$, $t \mapsto P_t\varphi(x)$ is continuous;
- $P_{t+s} = P_t P_s$, and $P_0 = Id_E$.

Remark 2.2.2. Notice that if $(\varphi_n)_{n\in\mathbb{N}} \subset C_b(E)$ is a sequence such that $\varphi_n \stackrel{\pi}{\to} \varphi \in C_b(E)$ as $n \to \infty$, then $P_t\varphi_n \stackrel{\pi}{\to} P_t\varphi$ as $n \to \infty$, for any $t \geq 0$.

In [40] semigroups as in Definition 2.2.1 are called *transition π-semigroups*. We have the following:

Theorem 2.2.3. *Let $(P_t)_{t \geq 0}$ be a stochastically continuous Markov semigroup. Then the family of linear maps $P_t^* : (C_b(E))^* \to (C_b(E))^*$, $t \geq 0$, defined by the formula*

$$\langle \varphi, P_t^* F \rangle_{\mathcal{L}(C_b(E),\,(C_b(E))^*)} = \langle P_t \varphi, F \rangle_{\mathcal{L}(C_b(E),\,(C_b(E))^*)}, \qquad (2.1)$$

where $t \geq 0$, $F \in (C_b(E))^*$, $\varphi \in C_b(E)$, is a semigroup of linear maps on $(C_b(E))^*$ of norm 1 and maps $\mathcal{M}(E)$ into $\mathcal{M}(E)$.

Proof. Clearly, P_t^* is linear. Let $F \in \big(C_b(E)\big)^*$, $t \geq 0$. We have, for any $\varphi \in C_b(E)$,

$$\langle \varphi, P_t^* F \rangle_{\mathcal{L}(C_b(E),\,(C_b(E))^*)} \leq \|\varphi\|_0 \|F\|_{(C_b(E))^*}.$$

Then $P_t : (C_b(E))^* \to (C_b(E))^*$ has norm equal to 1. Moreover, by (2.1) it follows easily that $P_t^*(P_s^* F) = P_{t+s}^* F$, for any $t, s \geq 0$, $F \in (C_b(E))^*$. Hence, (2.1) defines a semigroups of application in $(C_b(E))^*$ of norm equal to 1.

Now we prove that $P_t^* : \mathcal{M}(E) \to \mathcal{M}(E)$. According to Definition 2.2.1, let $\{\pi_t(x, \cdot), \ x \in E\}$ be the family of probability measures associated to P_t, that is $P_t \varphi(x) = \int_E \varphi(y) \pi_t(x, dy)$, for any $\varphi \in C_b(E)$. Note that that $P_t \varphi \geq 0$ for any $\varphi \geq 0$. This implies that if $\langle \varphi, F \rangle \geq 0$ for any $\varphi \geq 0$, then $\langle \varphi, P_t^* F \rangle \geq 0$ for any $\varphi \geq 0$. Hence, in order to check that $P_t^* : \mathcal{M}(E) \to \mathcal{M}(E)$, it is sufficient to take μ positive. So, let $\mu \in \mathcal{M}(E)$ be positive and consider the map

$$\Lambda : \mathcal{B}(E) \to \mathbb{R}, \qquad \Gamma \mapsto \Lambda(\Gamma) = \int_E \pi_t(x, \Gamma) \mu(dx).$$

Since for any $\Gamma \in \mathcal{B}(E)$ the map $E \to [0, 1]$, $x \to \pi_t(x, \Gamma)$ is Borel, the above formula in meaningful. It is easy to see that Λ is a positive and finite Borel measure on E, namely $\Lambda \in \mathcal{M}(E)$. Let us show $\Lambda = P_t^* \mu$.

Let us fix $\varphi \in C_b(E)$, and consider a sequence of simple Borel functions $(\varphi_n)_{n \in \mathbb{N}}$ which converges uniformly to φ and such that $|\varphi_n(x)| \leq |\varphi(x)|$, $x \in E$. For any $x \in E$ we have

$$\lim_{n \to \infty} \int_E \varphi_n(y) \pi_t(x, dy) = \int_E \varphi(y) \pi_t(x, dy) = P_t \varphi(x)$$

and

$$\sup_{x \in E} \left| \int_E \varphi_n(y) \pi_t(x, dy) \right| \leq \|\varphi\|_0.$$

Hence, by the dominated convergence theorem and by taking into account that φ_n is simple, we have

$$
\begin{aligned}
\int_E \varphi(x) \Lambda(dx) &= \lim_{n \to \infty} \int_E \varphi_n(x) \Lambda(dx) \\
&= \lim_{n \to \infty} \int_E \left(\int_E \varphi_n(y) \pi_t(x, dy) \right) \mu(dx) \\
&= \int_E \left(\int_E \varphi(y) \pi_t(x, dy) \right) \mu(dx) = \int_E P_t \varphi(x) \mu(dx).
\end{aligned}
$$

This implies that $P_t^* \mu = \Lambda$ and consequently $P_t^* \mu \in \mathcal{M}(E)$. $\qquad \square$

2.2.1. The infinitesimal generator

It is clear that a stochastically continuous Markov semigroup is not, in general, strongly continuous. However, we can define an infinitesimal generator $(K, D(K))$ by setting

$$
\begin{cases}
D(K) = \left\{ \varphi \in C_b(E) : \exists g \in C_b(E), \lim_{t \to 0^+} \dfrac{P_t \varphi(x) - \varphi(x)}{t} = g(x), \right. \\
\qquad \left. \forall x \in E, \quad \sup_{t \in (0,1)} \left\| \dfrac{P_t \varphi - \varphi}{t} \right\|_0 < \infty \right\} \\
K\varphi(x) = \lim_{t \to 0^+} \dfrac{P_t \varphi(x) - \varphi(x)}{t}, \quad \varphi \in D(K), \ x \in E.
\end{cases}
\tag{2.2}
$$

The next result is proved in Propositions 3.2, 3.3, 3.4 of [40]. For the reader's convenience, we give the complete proof.

Theorem 2.2.4. *Let us assume that $(P_t)_{t \geq 0}$ is as in Definition 2.2.1 and let $(K, D(K))$ be its infinitesimal generator, defined as in (2.2). Then*

(i) *for any $\varphi \in D(K)$, $P_t \varphi \in D(K)$ and $K P_t \varphi = P_t K \varphi$, $t \geq 0$;*
(ii) *for any $\varphi \in D(K)$, $x \in E$, the map $[0, \infty) \to \mathbb{R}$, $t \mapsto P_t \varphi(x)$ is continuously differentiable and $(d/dt) P_t \varphi(x) = P_t K \varphi(x)$;*
(iii) *for any $f \in C_b(E)$, $t > 0$ the map $E \to \mathbb{R}$, $x \mapsto \int_0^t P_s f(x) ds$ belongs to $D(K)$ and it holds*

$$
K \left(\int_0^t P_s f \, ds \right) = P_t f - f.
$$

Moreover, if $\varphi \in D(K)$ we have

$$
K \left(\int_0^t P_s f \, ds \right) = \int_0^t K P_s f \, ds;
$$

(iv) K *is a* π-*closed operator on* $C_b(E)$, *that is for any sequence* $\{\varphi_n\}_{n\in\mathbb{N}} \subset C_b(E)$ *such that* $\varphi_n \overset{\pi}{\to} \varphi \in C_b(E)$ *and* $K\varphi_n \overset{\pi}{\to} g \in C_b(E)$ *as* $n \to \infty$ *it follows that* $\varphi \in D(K)$ *and* $g = K\varphi$;

(v) $D(K)$ *is* π-*dense in* $C_b(E)$;

(vi) *for any* $\lambda > 0$ *the linear operator* $R(\lambda, K)$ *on* $C_b(E)$ *defined by*

$$R(\lambda, K)f(x) = \int_0^\infty e^{-\lambda t} P_t f(x)dt, \quad f \in C_b(E), \ x \in E$$

satisfies, for any $f \in C_b(E)$

$$R(\lambda, K) \in \mathcal{L}(C_b(E)), \qquad \|R(\lambda, K)\|_{\mathcal{L}(C_b(E))} \le \frac{1}{\lambda}$$

$$R(\lambda, K)f \in D(K), \qquad (\lambda I - K)R(\lambda, K)f = f.$$

We call $R(\lambda, K)$ *the* resolvent *of* K *at* λ.

Proof. (i). Take $\varphi \in D(K)$. Since $\varphi \in D(K)$ we have that

$$\lim_{h\to 0^+} \frac{P_h\varphi - \varphi}{h} \overset{\pi}{=} K\varphi.$$

Hence, by Remark 2.2.2

$$K P_t\varphi \overset{\pi}{=} \lim_{h\to 0^+} \frac{P_h P_t\varphi - P_t\varphi}{h} \overset{\pi}{=} \lim_{h\to 0^+} P_t\left(\frac{P_h\varphi - \varphi}{h}\right)$$

$$\overset{\pi}{=} P_t\left(\frac{P_h\varphi - \varphi}{h}\right) = P_t K\varphi.$$

(ii). By (i) we have $K P_t\varphi(x) = d/dt\, P_t\varphi(x) = P_t K\varphi(x)$. Since the map $t \mapsto P_t K\varphi(x)$ is continuous, (ii) follows.

(iii) First, we have to check that $\int_0^t P_s f ds$ belongs to $C_b(E)$. For any $x \in E$ we have

$$\left|\int_0^t P_s\varphi(x)ds\right| \le t\|\varphi\|_0.$$

Now let us fix $\varepsilon > 0$ and take $\delta > 0$ such that

$$\sup_{s\in[0,t]} |P_s\varphi(x) - P_s\varphi(y)| < \frac{\varepsilon}{t}$$

when $|x - y| < \delta$. This is possible since $(P_t)_{t\ge 0}$ is a locally bounded in $\mathcal{L}(C_b(E))$. Therefore, for any $x, y \in E$, $|x - y| < \delta$ we have

$$\left|\int_0^t P_s\varphi(x)ds - \int_0^t P_s\varphi(x)ds\right| \le \int_0^t |P_t\varphi(x) - P_t\varphi(y)|\, ds < \varepsilon.$$

Hence, $\int_0^t P_s\varphi ds \in C_b(E)$. Let us show $\int_0^t P_s\varphi ds \in D(K)$. Now, taking into account that for any $x \in E$ the integral $\int_0^t P_s\varphi(x)ds$ is a Riemann integral, we have

$$P_h\left(\int_0^t P_s\varphi ds\right) = \int_0^t P_{t+h}\varphi ds,$$

for any $h \geq 0$, since $\int_0^t P_s\varphi(x)ds$ is the limit with respect to the π-convergence of functions in $C_b(H)$. So, for any $x \in E, h > 0$ we have

$$P_h\left(\int_0^t P_s\varphi ds\right)(x) - \int_0^t P_s\varphi(x)ds = \int_0^t P_{t+h}\varphi(x)ds - \int_0^t P_s\varphi(x)ds$$

$$= \int_h^{t+h} P_s\varphi(x)ds - \int_0^t P_s\varphi(x)ds = \int_t^{t+h} P_s\varphi(x)ds - \int_0^h P_s\varphi(x)ds.$$

Therefore, by the continuity of $s \to P_s\varphi(x)$ we have

$$\lim_{h\to 0^+} \frac{1}{h}\left(P_h\left(\int_0^t P_s\varphi ds\right)(x) - \int_0^t P_s\varphi(x)ds\right) = P_t\varphi(x) - \varphi(x).$$

Finally, since $\|P_s\|_{\mathcal{L}(C_b(E))} \leq 1$ we find

$$\left|\frac{1}{h}\left(P_h\left(\int_0^t P_s\varphi ds\right)(x) - \int_0^t P_s\varphi(x)ds\right)\right| \leq 2\|\varphi\|_0.$$

This implies

$$\sup_{h\in(0,1)}\left\|\frac{1}{h}\left(P_h\left(\int_0^t P_s\varphi ds\right)(x) - \int_0^t P_s\varphi(x)ds\right)\right\|_0 < \infty.$$

This prove the first part of (v). Now take $\varphi \in D(K)$. By (ii) we see that for any $x \in E$ it holds

$$P_t\varphi(x) - \varphi(x) = \int_0^t \frac{d}{ds}P_s\varphi(x)ds = \int_0^t P_s K\varphi(x)ds.$$

Hence, (iii) follows.

(iv) Take $(\varphi_n)_{n\in\mathbb{N}} \subset D(K)$ such that $\varphi_n \xrightarrow{\pi} 0$ as $n \to \infty$ and $K\varphi_n \xrightarrow{\pi} g \in C_b(H)$ as $n \to \infty$. By (iii) and Remark 2.2.2, for any $t > 0$ we have

$$P_t\varphi - \varphi \stackrel{\pi}{=} \lim_{n\to\infty}(P_t\varphi_n - \varphi_n)$$

$$\stackrel{\pi}{=} \lim_{n\to\infty}\int_0^t P_s K\varphi_n ds \stackrel{\pi}{=} \int_0^t P_s g ds.$$

Hence it follows easily

$$\lim_{t \to 0^+} \frac{P_t \varphi - \varphi}{t} \overset{\pi}{=} \lim_{t \to 0^+} \frac{1}{t} \int_0^t P_s g \, ds \overset{\pi}{=} g,$$

which implies $\varphi \in D(K)$ and $K\varphi = g$.

(v). Take $\varphi \in C_b(E)$ and set

$$\varphi_n = n \int_0^{\frac{1}{n}} P_s \varphi \, ds.$$

By (iii) we have $\varphi \in D(K)$. Since for any $x \in E$ the map $[0, \infty) \to \mathbb{R}$, $s \mapsto P_s \varphi(x)$ is continuous, we have $|\varphi_n(x) - \varphi(x)|_E \to 0$ as $n \to \infty$, for any $x \in E$. Finally, we have $|\varphi_n(x)| \leq \|\varphi\|_0$, which implies $\varphi_n \overset{\pi}{\to} \varphi$ as $n \to \infty$.

(vi) For any $\lambda > 0$ and for any $f \in C_b(E)$ we set

$$F_\lambda f(x) = \int_0^\infty e^{-\lambda t} P_t f(x) dt, \quad x \in E.$$

Fix $\lambda > 0$ and for any $f \in C_b(E)$.

Step 1. We first prove that $F_\lambda f \in C_b(E)$. Notice that for any $f \in C_b(E)$, $\lambda > 0$ the integral on the right-hand side of is meaningful, since $P_t f(x)$ is continuous and

$$\left| e^{-\lambda t} P_t f(x) \right| \leq e^{-\lambda t} \|f\|_0.$$

Hence,

$$\sup_{x \in E} |F_\lambda f(x)| \leq \frac{\|f\|_0}{\lambda}.$$

for any $f \in C(E)$. To prove the claim, it is sufficient to check that the function $E \to \mathbb{R}$, $x \mapsto F_\lambda f(x)$ is uniformly continuous. So, let us fix $\varepsilon > 0$. Let $T > 0$ such that

$$\left(\frac{2\|f\|_0}{\lambda} \right) e^{-\lambda T} < \frac{\varepsilon}{2}. \tag{2.3}$$

There exists $\delta > 0$, depending on f and T, such that if $x, y \in E$ and $|x - y|_E < \delta$ we have

$$\sup_{t \in [0,T]} |P_t f(x) - P_t f(y)| < \frac{\varepsilon}{2} \left(\frac{\lambda}{1 - e^{-\lambda T}} \right). \tag{2.4}$$

Then, if $x, y \in E$ and $|x - y|_E < \delta$ it holds

$$|F_\lambda f(x) - F_\lambda f(y)|$$

$$\leq \int_0^T e^{-\lambda t} |P_t f(x) - P_t f(y)| \, dt + \int_T^\infty e^{-\lambda t} |P_t f(x) - P_t f(y)| \, dt$$

$$\leq \frac{\varepsilon}{2} \left(\frac{\lambda}{1 - e^{-\lambda T}} \right) \int_0^T e^{-\lambda t} dt + 2\|f\|_0 \int_T^\infty e^{-\lambda t} dt < \varepsilon$$

thanks to (2.3), (2.4).

Step 2. Here we are checking $F_\lambda f \in D(K)$ and $(\lambda - K)F_\lambda f = f$. Set $g = F_\lambda f$ and $g_T = \int_0^T e^{-\lambda t} P_t f(x) dt$. We have

$$\lim_{T \to \infty} \|g - g_T\|_0 \leq \|f\|_0 \int_T^\infty e^{-\lambda t} dt = 0.$$

For any $h > 0$, $x \in E$ we have

$$\frac{P_h g(x) - g(x)}{h} = \frac{1}{h} \int_0^T e^{-\lambda t} \left(P_{t+h} f(x) - P_t f(x) \right) dt$$

$$= \Gamma_1(h, x) + \Gamma_2(h, x),$$

where

$$\Gamma_1(h, x) = \frac{e^{\lambda h} - 1}{h} g(x),$$

$$\Gamma_2(h, x) = \frac{e^{\lambda h}}{h} \int_0^h e^{-\lambda t} P_t f(x) dt.$$

We clearly have

$$\lim_{h \to 0^+} \Gamma_1(h, x) = \lambda g(x), \quad x \in H$$

and

$$\sup_{h \in (0,1)} \|\Gamma_1(h, \cdot)\|_0 \leq \sup_{h \in (0,1)} \left\{ \frac{e^{\lambda h} - 1}{h} \right\} \|f\|_0 \int_0^\infty e^{-\lambda t} dt < \infty.$$

Hence, $\Gamma_1(h, \cdot) \xrightarrow{\pi} \lambda g$ as $h \to 0^+$. Concerning the term $\Gamma_2(h, x)$, for any $x \in E$ we have

$$\lim_{h \to 0^+} \Gamma_2(h, x) = f(x),$$

since the map $[0, \infty) \to \mathbb{R}, t \mapsto e^{-\lambda t} P_t f(x)$ is continuous. On the other hand we have

$$|\Gamma_2(h, x)| \leq \frac{e^{\lambda h}}{h} \int_0^h e^{-\lambda t} dt \|f\|_0 = \frac{e^{\lambda h} (1 - e^{-\lambda h})}{\lambda h} \|f\|_0$$

which implies

$$\sup_{h \in (0,1)} \|\Gamma_2(h, \cdot)\|_0 < \infty.$$

Hence, $\Gamma_2(h, \cdot) \xrightarrow{\pi} f$ as $h \to 0^+$. We have found that $F_\lambda \in D(K)$ and that $K F_\lambda f = \lambda F_\lambda f + f$. This implies $(\lambda - K) F_\lambda f = f$. $\square$

Definition 2.2.5. We shall say that a set $D \subset D(K)$ is a π-core for the operator $(K, D(K))$ if D is π-dense in $C_b(E)$ and for any $\varphi \in D(K)$ there exists $m \in \mathbb{N}$ and an m-indexed sequence $\{\varphi_{n_1,\ldots,n_m}\}_{n_1 \in \mathbb{N}, \ldots, n_m \in \mathbb{N}} \subset D$ such that

$$\lim_{n_1 \to \infty} \cdots \lim_{n_m \to \infty} \varphi_{n_1,\ldots,n_m} \overset{\pi}{=} \varphi$$

and

$$\lim_{n_1 \to \infty} \cdots \lim_{n_m \to \infty} K \varphi_{n_1,\ldots,n_m} \overset{\pi}{=} K\varphi.$$

It is clear that a π-core is nothing but the extension of the notion of *core* with respect to the π-convergence. A useful example of core is given by the following

Proposition 2.2.6. *Let $(P_t)_{t \geq 0}$ be a stochastically continuous Markov semigroup and let $(K, D(K))$ be its infinitesimal generator. If $D \subset D(K)$ is π-dense in $C_b(E)$ and $P_t(D) \subset D$ for all $t \geq 0$, then D is a π-core for $(K, D(K))$.*

Proof. In order to get the result, we proceed as in [28]. Let $\varphi \in D(K)$. Since D in π-dense in $C_b(E)$, there exists a sequence $(\varphi_{n_2})_{n_2 \in \mathbb{N}} \subset D$ (for the sack of simplicity we assume that the sequence has only one index) such that $\varphi_{n_2} \xrightarrow{\pi} \varphi$ as $n_2 \to \infty$. Set

$$\varphi_{n_1,n_2,n_3}(x) = \frac{1}{n_3} \sum_{i=1}^{n_3} P_{\frac{i}{n_1 n_3}} \varphi_{n_2}(x) \tag{2.5}$$

for any $n_1, n_2, n_3 \in \mathbb{N}$. By Hypothesis, $(\varphi_{n_1,n_2,n_3}) \subset D$. Taking into account Remark 2.2.2, a straightforward computation shows that for any $x \in E$

$$\lim_{n_1 \to \infty} \lim_{n_2 \to \infty} \lim_{n_3 \to \infty} \varphi_{n_1,n_2,n_3}(x) = \lim_{n_1 \to \infty} \lim_{n_2 \to \infty} n_1 \int_0^{\frac{1}{n_1}} P_t \varphi_{n_2}(x) dt$$

$$= \lim_{n_1 \to \infty} n_1 \int_0^{\frac{1}{n_1}} P_t \varphi(x) dt = \varphi(x).$$

Moreover,

$$\sup_{n_1,n_2,n_3\in\mathbb{N}} \|\varphi_{n_1,n_2,n_3}\|_0 \le \sup_{n_2} \|\varphi_{n_2}\|_0 < \infty$$

since $\varphi_{n_2} \xrightarrow{\pi} \varphi$ as $n_2 \to \infty$. Hence,

$$\lim_{n_1\to\infty} \lim_{n_2\to\infty} \lim_{n_3\to\infty} \varphi_{n_1,n_2,n_3} \overset{\pi}{=} \varphi.$$

Similarly, since $D \subset D(K)$ and Theorem 2.2.4 holds, we have

$$
\begin{aligned}
\lim_{n_3\to\infty} K\varphi_{n_1,n_2,n_3}(x) &= n_1 \int_0^{\frac{1}{n_1}} K P_t \varphi_{n_2}(x)dt \\
&= n_1\left(P_{\frac{1}{n_1}} \varphi_{n_2}(x) - \varphi_{n_2}(x) \right).
\end{aligned}
$$

So we find

$$
\lim_{n_1\to\infty} \lim_{n_2\to\infty} \lim_{n_3\to\infty} K\varphi_{n_1,n_2,n_3}(x) = \lim_{n_1\to\infty} \lim_{n_2\to\infty} n_1\left(P_{\frac{1}{n_1}} \varphi_{n_2}(x) - \varphi_{n_2}(x) \right)
$$

$$
= \lim_{n_1\to\infty} n_1\left(P_{\frac{1}{n_1}} \varphi(x) - \varphi(x) \right) = K\varphi(x), \qquad (2.6)
$$

since $\varphi \in D(K)$. To conclude the proof, we have to show that these limits are uniformly bounded with respect to every index. Indeed we have

$$\sup_{n_3\in\mathbb{N}} \|K\varphi_{n_1,n_2,n_3}\| \le \|K\varphi_{n_2}\| < \infty,$$

$$\sup_{n_2\in\mathbb{N}} \left\| n_1\left(P_{\frac{1}{n_1}} \varphi_{n_2} - \varphi_{n_2} \right) \right\|_0 \le 2n_1 \sup_{n_2\in\mathbb{N}} \|\varphi_{n_2}\|_0 < \infty.$$

Finally, the last limit in (2.6) is uniformly bounded with respect to n_1 since $\varphi \in D(K)$. $\qquad\square$

2.3. The measure equation

Theorem 2.3.1. *Let $(P_t)_{t\ge 0}$ be a stochastically continuous Markov semi-group, as in Definition 2.2.1, and let $(K, D(K))$ be its infinitesimal generator in $C_b(E)$, defined by (2.2). Then, for any $\mu \in \mathcal{M}(E)$ there exists a unique family of measures $\{\mu_t, \ t \ge 0\} \subset \mathcal{M}(E)$ fulfilling*

$$\int_0^T |\mu_t|_{TV}(E)dt < \infty, \qquad T > 0; \qquad (2.7)$$

$$\int_E \varphi(x)\mu_t(dx) - \int_E \varphi(x)\mu(dx) = \int_0^t \left(\int_E K\varphi(x)\mu_s(dx) \right)ds, \quad (2.8)$$

for any $\varphi \in D(K)$, $t \ge 0$, and the solution is given by $P_t^*\mu$, $t \ge 0$.

Proof. Let us fix $\mu \in \mathcal{M}(E)$. We split the proof into two parts.

2.3.1. Existence of a solution

By Theorem 2.2.3, formula (2.1) defines a family $\{P_t^*\mu, ; t \geq 0\}$ of finite Borel measures on E. Moreover, $\|P_t^*\mu\|_{(C_b(E))^*} \leq \|\mu\|_{(C_b(E))^*} = |\mu|_{TV}(E)$. Hence, (2.7) follows. Since for any $\varphi \in C_b(E)$ it holds

$$\lim_{t \to 0^+} \int_E P_t \varphi(x) \mu(dx) = \int_E \varphi(x) \mu(dx),$$

by the semigroup property of P_t it follows that for any $\varphi \in C_b(E)$ the function

$$\mathbb{R}^+ \to \mathbb{R}, \quad t \mapsto \int_E \varphi(x) P_t^* \mu(dx) \tag{2.9}$$

is continuous. Clearly, $P_0^*\mu = \mu$. Now we show that if $\varphi \in D(K)$ then the function (2.9) is differentiable. Indeed, by taking into account (2.2) and that $P_t^*\mu \in \mathcal{M}(E)$, we can apply the dominated convergence theorem for any $\varphi \in D(K)$ to obtain

$$\frac{d}{dt} \int_E \varphi(x) P_t^* \mu(dx)$$

$$= \lim_{h \to 0} \frac{1}{h} \left(\int_E P_{t+h} \varphi(x) \mu(dx) - \int_E P_t \varphi(x) \mu_t(dx) \right)$$

$$= \lim_{h \to 0} \int_E \left(\frac{P_{t+h}\varphi(x) - P_t\varphi(x)}{h} \right) \mu(dx)$$

$$= \lim_{h \to 0} \int_E P_t \left(\frac{P_h \varphi - \varphi}{h} \right)(x) \mu(dx)$$

$$= \int_E \lim_{h \to 0} \left(\frac{P_h \varphi - \varphi}{h} \right)(x) P_t^* \mu(dx) = \int_E K\varphi(x) P_t^* \mu(dx).$$

Then, by arguing as above, the differential of (2.9) is continuous. This clearly implies that $\{P_t^*\mu, \ t \geq 0\}$ is a solution of the measure equation (2.8).

2.3.2. Uniqueness of the solution

Since problem (2.8) is linear, it is enough to take $\mu = 0$. We claim that in this case $\mu_t = 0$, $\forall t \geq 0$. In order to prove this, let us fix $T > 0$ and let us consider the Kolmogorov backward equation

$$\begin{cases} u_t(t, x) + Ku(t, x) = \varphi(x) & t \in [0, T], \ x \in E, \\ u(T, x) = 0, \end{cases} \tag{2.10}$$

where $\varphi \in C_b(E)$. The meaning of (2.10) is explained by the following lemma.

Lemma 2.3.2. *For any $T > 0$, $\varphi \in C_b(E)$ the real valued function*

$$u : [0, T] \times E \to \mathbb{R}$$

$$u(t, x) = - \int_0^{T-t} P_s \varphi(x) ds, \quad (t, x) \in [0, T] \times E. \quad (2.11)$$

satisfies the following statements

(i) $u \in C_b([0, T] \times E)$ [1];
(ii) $u(t, \cdot) \in D(K)$ *for any* $t \in [0, T]$ *and the function* $[0, T] \times E \to \mathbb{R}$, $(t, x) \mapsto Ku(t, x)$ *is continuous and bounded;*
(iii) *the real valued function* $[0, T] \times E \to \mathbb{R}$, $(t, x) \mapsto u(t, x)$ *is differentiable with respect to t with continuous and bounded differential $u_t(t, x)$, and the function* $[0, T] \times E \to \mathbb{R}$, $(t, x) \mapsto u_t(t, x)$ *is continuous and bounded;*
(iv) *for any* $(t, x) \in [0, T] \times E$ *the function u satisfies* (2.10).

Proof. For any $s, t \in [0, T]$, $s \le t$ we have

$$
\begin{aligned}
u(t, x) - u(s, x) &= - \int_0^{T-t} P_\tau \varphi(x) d\tau + \int_0^{T-s} P_\tau \varphi(x) d\tau \\
&= \int_{T-t}^{T-s} P_\tau \varphi(x) d\tau.
\end{aligned}
$$

Then

$$\|u(t, \cdot) - u(s, \cdot)\|_0 \le |t - s| \|\varphi\|_0.$$

(i) is proved. By (vi) of Theorem 2.2.4, $u(t, \cdot) \in D(K)$ for any $t \in [0, T]$ and it holds $Ku(t, x) = -P_{T-t}\varphi(x) + \varphi(x)$, for any $x \in E$. So (ii) follows (*cf.* Definition 2.2.1). Now let $h \in (-t, T - t)$ and $x \in E$. We have

$$\frac{u(t + h, x) - u(t, x)}{h} + Ku(t, x) - \varphi(x) \quad (2.12)$$

$$= \frac{1}{h} \int_{T-t-h}^{T-t} P_\tau \varphi(x) ds - P_{T-t}\varphi(x)$$

Then, since $P_t \varphi(x)$ is continuous in t, (2.12) vanishes as $h \to 0$. This implies that $u(t, x)$ is differentiable with respect to t and (2.10) holds. Moreover, by (ii), we have that the map $t \mapsto u_t(t, x) = -Ku(t, x) + \varphi(x)$ is continuous. This proves (iii) and (iv). The proof is complete. $\square$

---

[1] Clearly, $C_b([0, T] \times E)$ is isomorphic to $C([0, T]; C_b(E))$

We need the following:

Lemma 2.3.3. *Let $\{\mu_t\}$ be a solution of the measure equation* (2.7), (2.8).
Then, for any function $u : [0, T] \times E \to \mathbb{R}$ satisfying statements (i), (ii),
(iii) *of Lemma* 2.3.2 *the map*

$$[0, T] \to \mathbb{R}, \qquad t \mapsto \int_E u(t, x)\mu_t(dx)$$

is absolutely continuous and for any $t \geq 0$ it holds

$$\int_E u(t, x)\mu_t(dx) - \int_E u(0, x)\mu(dx)$$
$$= \int_0^t \left(\int_E \big(u_s(s, x) + Ku(s, x)\big)\mu_s(dx) \right) ds. \quad (2.13)$$

Proof. We split the proof in several steps.

Step 1. *Approximation of $u(t, x)$.*
With no loss of generality, we assume $T = 1$. For any $x \in E$, let us
consider the approximating functions $\{u^n(\cdot, x)\}_{n\in\mathbb{N}}$ of $u(\cdot, x)$ given by
the Bernstein polynomials (see, for instance, [46], section 0.2). Namely,
for any $n \in \mathbb{N}$, $x \in E$ we consider the function

$$[0, T] \to \mathbb{R}, \quad t \mapsto u^n(t, x) = \sum_{k=0}^n \alpha_{k,n}(t)u\left(\frac{k}{n}, x\right),$$

where

$$\alpha_{k,n}(t) = \binom{n}{k}t^k(1 - t)^{n-k}.$$

Since $u \in C([0, T]; C_b(E))$, it is well known that it holds

$$\lim_{n\to\infty} \sup_{t\in[0,1]} \|u^n(t, \cdot) - u(t, \cdot)\|_0 = 0 \qquad (2.14)$$

and

$$\sup_{t\in[0,1]} \|u^n(t, \cdot)\|_0 < \infty, \quad n \in \mathbb{N}.$$

Then, for any $t \in [0, 1]$

$$\lim_{n\to\infty} u^n(t, \cdot) \overset{\pi}{=} u(t, \cdot). \qquad (2.15)$$

We also have that for any $n \in \mathbb{N}$, $t \in [0, 1]$

$$u^n(t, \cdot) \in D(K),$$

and that for any $x \in E$ the function $[0, 1] \to \mathbb{R}$, $t \mapsto Ku^n(t, x)$ is continuous (*cf.* (ii) of Lemma 2.3.2). Then, for any $x \in E$ it holds

$$\lim_{n \to \infty} \sup_{t \in [0,1]} |Ku^n(t, x) - Ku(t, x)| = 0,$$

$$\sup_{t \in [0,1]} \|Ku^n(t, \cdot)\|_0 \leq \sup_{t \in [0,1]} \|Ku(t, \cdot)\|_0 < \infty. \qquad (2.16)$$

This clearly implies that for any $t \in [0, 1]$

$$\lim_{n \to \infty} Ku^n(t, \cdot) \overset{\pi}{=} Ku(t, \cdot). \qquad (2.17)$$

Similarly, since for any x the function $t \mapsto u(t, x)$ is differentiable with respect to t, we also have that for any $x \in E$

$$\lim_{n \to \infty} \sup_{t \in [0,1]} |u_t^n(t, x) - u_t(t, x)| = 0,$$

$$\sup_{t \in [0,1]} \|u_t^n(t, \cdot)\|_0 \leq \sup_{t \in [0,1]} \|u_t(t, \cdot)\|_0 < \infty. \qquad (2.18)$$

Hence, for any $t \in [0, 1]$

$$\lim_{n \to \infty} u_t^n(t, \cdot) \overset{\pi}{=} u_t(t, \cdot). \qquad (2.19)$$

Step 2. *Differential of $\int_E u^n(t, x) \mu_t(dx)$.*
For any $n \in \mathbb{N}$, $k \leq n$ and for almost all $t \in [0, 1]$ we have

$$\frac{d}{dt} \left(\int_E \alpha_{k,n}(t) u\left(\frac{k}{n}, x\right) \mu_t(dx) \right)$$

$$= \frac{d}{dt} \left(\alpha_{k,n}(t) \int_E u\left(\frac{k}{n}, x\right) \mu_t(dx) \right)$$

$$= \alpha'_{k,n}(t) \int_E u\left(\frac{k}{n}, x\right) \mu_t(dx) + \alpha_{k,n}(t) \int_E Ku\left(\frac{k}{n}, x\right) \mu_t(dx).$$

$$= \int_E \left(\alpha'_{k,n}(t) u\left(\frac{k}{n}, x\right) + \alpha_{k,n}(t) Ku\left(\frac{k}{n}, x\right) \right) \mu_t(dx).$$

Note that the last terms belong to $L^1([0, 1])$. This implies

$$\int_E u^n(t, x) \mu_t(dx) - \int_E u^n(0, x) \mu(dx)$$

$$= \int_0^t \left(\int_E \left(u_s^n(s, x) + Ku^n(s, x) \right) \mu_s(dx) \right) ds,$$

for any $n \in \mathbb{N}$.

Step 3. *Conclusion.*
Consider the functions

$$f : [0, 1] \to \mathbb{R}, \quad f(t) = \int_E u(t, x)\mu_t(dx)$$

and

$$f_n : [0, 1] \to \mathbb{R}, \quad f_n(t) = \int_E u^n(t, x)\mu_t(dx).$$

By 2.14 we have

$$\left| \int_E \left(u^n(t, x) - u(t, x) \right)\mu_t(dx) \right| \leq \sup_{t\in[0,1]} \|u^n(t, \cdot) - u(t, \cdot)\|_0 |\mu_t|_{TV}(E).$$

Since (2.7) and (2.14) hold, it follows that the sequence $(f_n)_{n\in\mathbb{N}}$ converges to f in $L^1([0, 1])$, as $n \to \infty$. We also have, by Step 2, that f_n is absolutely continuous and hence differentiable for almost all $t \in [0, 1]$, with differential in $L^1([0, 1])$ given by

$$f_n'(t) = \int_E \left(u_t^n(t, x) + Ku^n(t, x) \right)\mu_t(dx),$$

for almost all $t \in [0, 1]$. By (2.17), (2.19) we have

$$
\begin{aligned}
\lim_{n\to\infty} f_n'(t) &= \lim_{n\to\infty} \int_E \left(u_t^n(t, x) + Ku^n(t, x) \right)\mu_t(dx) \\
&= \int_E \left(u_t(t, x) + Ku(t, x) \right)\mu_t(dx), \quad\quad (2.20)
\end{aligned}
$$

for all $t \in [0, T]$. Moreover, it holds

$$\sup_{n\in\mathbb{N}} |f_n'(t)| \leq \left(\sup_{t\in[0,1]} \|u(t, \cdot)\|_0 + \sup_{t\in[0,1]} \|Ku(t, \cdot)\| \right) |\mu_t|_{TV}(E).$$

Hence, still by (2.16) and (2.18), there exists a constant $c > 0$ such that $\sup_n |f_n'(t)| \leq c|\mu_t|_{TV}(E)$. By taking into account (2.7), it follows that the limit in (2.20) holds in $L^1([0, 1])$. Let us denote by $g(t)$ the right-hand side of (2.20). We find, for any $a, b \in [0, 1]$,

$$
\begin{aligned}
f(b) - f(a) &= \lim_{n\to\infty} \left(f_n(b) - f_n(a) \right) \\
&= \lim_{n\to\infty} \int_a^b f_n'(t)dt = \int_a^b \lim_{n\to\infty} f_n'(t)dt = \int_a^b g(t)dt.
\end{aligned}
$$

Therefore, f is absolutely continuous, and $f'(t) = g(t)$ for almost all $t \in [0, 1]$. Lemma 2.3.3 is proved. $\qquad\square$

Now let $\varphi \in C_b(E)$ and let u be the function defined in (2.11). Then u satisfies statements (i)–(iv) of Lemma 2.3.2. Hence, by Lemma 2.3.3 it follows that the function $[0, T] \to \mathbb{R}, t \to \int_E u(t, x)\mu_t(dx)$ is absolutely continuous, with differential

$$\frac{d}{dt} \int_E u(t, x)\mu_t(dx) = \int_E \left(u_t(t, x) + Ku(t, x) \right)\mu_t(dx)$$
$$= \int_E \varphi(x)\mu_t(dx),$$

for almost all $t \in [0, T]$. So, we can write

$$0 = \int_E u(T, x)\mu_T(dx) - \int_E u(0, x)\mu(dx) =$$
$$= \int_0^T \left(\frac{d}{dt} \int_E u(t, x)\mu_t(dx) \right) dt$$
$$= \int_0^T \left(\int_E \varphi(x)\mu_t(dx) \right) dt.$$

for all $\varphi \in C_b(E)$. By the arbitrariness of T, it follows that for any $t \geq 0$ it holds

$$\int_0^t \left(\int_E \varphi(x)\mu_s(dx) \right) ds = 0.$$

In particular, the above identity holds true for $\varphi = K\psi$, for any $\psi \in D(K)$. Then, taking into account (2.8), it follows that for any $\psi \in D(K)$, $t \geq 0$ it holds

$$\int_E \psi(x)\mu_t(dx) = 0. \tag{2.21}$$

Finally, since $D(K)$ is π-dense in $C_b(E)$ (cf. (v) of Theorem 2.2.4), (2.21) holds for any $\psi \in C_b(E)$, $t \geq 0$ and consequently $\mu_t = 0$ for any $t \geq 0$. The proof is now complete. $\qquad \square$

Chapter 3
Measure equations for Ornstein-Uhlenbeck operators

3.1. Introduction and main results

We denote by H a separable Hilbert space with norm $|\cdot|$ and inner product $\langle \cdot, \cdot \rangle$ and we consider the stochastic differential equation in H

$$\begin{cases} dX(t) = AX(t)dt + BdW(t), & t \geq 0 \\ X(0) = x \in H, \end{cases} \qquad (3.1)$$

where A, B, $\{W(t)\}_{t\geq 0}$ fulfil Hypothesis 1.2.1. In the following, we set $Q = BB^*$. For any $x \in h$, equation (3.1) has a unique mild solution $X(t, x)$, $t \geq 0$, that is a square integrable random process adapted to the filtration $(\mathcal{F}_t)_{t\geq 0}$, given by

$$X(t, x) = e^{tA}x + W_A(t). \qquad (3.2)$$

It is well known that the random variable $X(t, x)$ has Gaussian law of mean $e^{tA}x$ and covariance operator Q_t (*cf.* Hypothesis 1.2.1). Hence, the corresponding transition semigroup $(R_t)_{t\geq 0}$, called the *Ornstein-Uhlenbeck* (in the following, OU) semigroup enjoys the representation

$$R_t\varphi(x) = \int_H \varphi(e^{tA}x + y)N_{Q_t}(dy), \qquad \varphi \in C_b(H), \ t \geq 0, \ x \in H, \quad (3.3)$$

where N_{Q_t} is the Gaussian measure on H of zero mean and covariance operator Q_t (see [23]). Of course, in the above formula we mean $N_{Q_0} = \delta_0$. Also, the OU semigroup $(R_t)_{t\geq 0}$ is a stochastically continuous Markov semigroup in $C_b(H)$. Moreover, it is well known that for any $t \geq 0$, $h \in H$ it holds[1]

$$R_t e^{i\langle \cdot, h\rangle}(x) = e^{i\langle e^{tA}x, h\rangle - \frac{1}{2}\langle Q_t h, h\rangle}, \qquad h \in H. \qquad (3.4)$$

---

[1] R_t acts on reals functions, but it can be trivially extended to complex ones.

We denote by $(L, D(L))$ the infinitesimal generator

$$
\begin{cases}
D(L) = \left\{ \varphi \in C_b(H) : \exists g \in C_b(H), \lim_{t \to 0^+} \dfrac{P_t \varphi(x) - \varphi(x)}{t} = g(x), \right. \\
\qquad\qquad \left. \forall x \in H, \ \sup_{t \in (0,1)} \left\| \dfrac{P_t \varphi - \varphi}{t} \right\|_0 < \infty \right\} \\
\\
L\varphi(x) = \lim_{t \to 0^+} \dfrac{P_t \varphi(x) - \varphi(x)}{t}, \qquad \varphi \in D(L), \ x \in H.
\end{cases}
\tag{3.5}
$$

By Theorem 2.3.1 follows

Theorem 3.1.1. *Let $(R_t)_{t \geq 0}$ be the Ornstein-Uhlenbeck semigroup (3.3), and let $(L, D(L))$ be its infinitesimal generator. Then, for any $\mu \in \mathcal{M}(H)$ there exists a unique family of measures $\{\mu_t, \ t \geq 0\} \subset \mathcal{M}(H)$ fulfilling*

$$
\int_0^T |\mu_t|_{TV}(H) dt < \infty, \qquad T > 0;
\tag{3.6}
$$

$$
\int_E \varphi(x) \mu_t(dx) - \int_H \varphi(x)\mu(dx) = \int_0^t \left(\int_H L\varphi(x)\mu_s(dx) \right) ds,
\tag{3.7}
$$

for any $\varphi \in D(L)$, $t \geq 0$. Moreover, $\mu_t = R_t^* \mu$, $t \geq 0$.

We are interested in extending the previous results to the Kolmogorov operator associated to equation (3.1), which looks like

$$
\frac{1}{2}\mathrm{Tr}[QD^2\varphi(x)] + \langle x, A^*D\varphi(x) \rangle, \qquad x \in H.
$$

To this purpose, we need some preliminary results. It will be helpful the following result about approximation of $C_b(H)$-functions.

Proposition 3.1.2. *We recall that $\mathcal{E}(H)$ is the linear span of the real and imaginary part of the functions*

$$
H \to \mathbb{C}, \qquad x \mapsto e^{i\langle x, h \rangle},
$$

where $h \in H$. Then $\mathcal{E}(H)$ is π-dense in $C_b(H)$ and for any $\varphi \in C_b(H)$ there exists a two-indexed sequence $(\varphi_{n_1, n_2}) \subset \mathcal{E}(H)$ such that

$$
\lim_{n_1 \to \infty} \lim_{n_2 \to \infty} \varphi_{n_1, n_2}(x) = \varphi(x), \ x \in H
\tag{3.8}
$$

$$
\sup_{n_1, n_2} \|\varphi_{n_1, n_2}\|_0 \leq \|\varphi\|_0.
\tag{3.9}
$$

Moreover, if $\varphi \in C_b^1(H)$ we can choose the sequence $(\varphi_{n_1,n_2}) \subset \mathcal{E}(H)$ in such a way that (3.8), (3.9) hold and for any $h \in H$

$$\lim_{n_1 \to \infty} \lim_{n_2 \to \infty} \langle D\varphi_{n_1,n_2}(x), h \rangle = \langle D\varphi(x), h \rangle, \quad x \in H$$

$$\sup_{n_1,n_2} \|D\varphi_{n_1,n_2}\|_{C_b(H;H)} \leq \|D\varphi\|_{C_b(H;H)}. \tag{3.10}$$

Proof. (3.8) and (3.9) are proved in [13], Proposition 1.2. (3.10) follows by the well known properties of the Fourier approximation with Fejér kernels of differentiable functions (see, for instance, [33]). □

We are going to improve this result. We recall that the set $\mathcal{E}_A(H)$ has been introduced in Section 1.1.

Proposition 3.1.3. *For any $\varphi \in C_b(H)$ there exists a three-indexed sequence $(\varphi_{n_1,n_2,n_3}) \subset \mathcal{E}_A(H)$ such that*

$$\lim_{n_1 \to \infty} \lim_{n_2 \to \infty} \lim_{n_3 \to \infty} \varphi_{n_1,n_2,n_3} \overset{\pi}{=} \varphi.$$

Moreover, if $\varphi \in C_b^1(H)$, we have that for any $h \in H$ it holds

$$\lim_{n_1 \to \infty} \lim_{n_2 \to \infty} \lim_{n_3 \to \infty} \langle D\varphi_{n_1,n_2,n_3}, h \rangle \overset{\pi}{=} \langle D\varphi, h \rangle.$$

Proof. Let $\varphi \in C_b(H)$ and let us consider a two-indexed sequence $(\varphi_{n_1,n_2}) \subset \mathcal{E}(H)$ as in Proposition 3.1.2. Let us define the sequence (φ_{n_1,n_2,n_3}) by setting

$$\varphi_{n_1,n_2,n_3}(x) = \varphi_{n_1,n_2}(n_3 R(n_3, A^*)x), \quad x \in H, \ n_3 \in \mathbb{N},$$

where $R(n_3, A^*)$ is the resolvent operator of A^* at n_3. Clearly, $\varphi_{n_1,n_2,n_3} \in \mathcal{E}_A(H)$. Taking into account that $nR(n, A^*)x \to x$ as $n \to \infty$ for all $x \in H$, and that for some $c > 0$ it holds $|nR(n, A^*)x| \leq c|x|$ for any $x \in H, n \geq 1$, it follows that $\varphi_{n_1,n_2,n_3} \overset{\pi}{\to} \varphi_{n_1,n_2}$ as $n_3 \to \infty$. If $f \in C_b^1(H)$, we observe that

$$\langle D(f(nR(n, A^*)\cdot))(x), h \rangle = \langle Df(nR(n, A^*)x), nR(n, A)h \rangle.$$

Therefore, be arguing as above, we find $\langle D(f(nR(n, A^*)\cdot)), h \rangle \overset{\pi}{\to} \langle Df(\cdot), h \rangle$ as $n \to \infty$. Hence the result follows. □

Example 3.1.4. If $A \neq 0$ we have $D(L) \cap \mathcal{E}_A(H) = \{\text{constant functions}\}$. In fact for any $x \in H, h \in D(A^*)$ we have

$$\lim_{t \to 0^+} \frac{R_t e^{i\langle h,x \rangle} - e^{i\langle h,x \rangle}}{t} = \left[-\frac{1}{2} \langle Qh, h \rangle + i\langle A^*h, x \rangle \right] e^{i\langle h,x \rangle},$$

which is not bounded when $h \neq 0$ and $A \neq 0$.

Let $\mathcal{I}_A(H)$ be the linear span of the real and imaginary part of the functions

$$H \to \mathbb{C}, \quad x \mapsto \int_0^a e^{i\langle e^{sA}x,h\rangle - \frac{1}{2}\langle Q_s h, h\rangle} ds : a > 0, \ h \in D(A^*),$$

where $D(A^*)$ is the domain of the adjoint operator of A.

Proposition 3.1.5. *The set $\mathcal{I}_A(H)$ is π-dense in $C_b(H)$, it is stable for R_t and $\mathcal{I}_A(H) \subset D(L)$. Moreover, it is a π-core for $(L, D(L))$ and for any $\varphi \in \mathcal{I}_A(H)$ it holds*

$$L\varphi(x) = \frac{1}{2}\mathrm{Tr}[QD^2\varphi(x)] + \langle x, A^*D\varphi(x)\rangle, \quad x \in H. \tag{3.11}$$

Proof. Let $h \in D(A^*)$ and $a > 0$. We have

$$\lim_{a \to 0^+} \frac{1}{a}\int_0^a e^{i\langle e^{sA}x,h\rangle - \frac{1}{2}\langle Q_s h, h\rangle} ds = e^{i\langle x,h\rangle}, \quad x \in H$$

and

$$\sup_{a>0}\left|\frac{1}{a}\int_0^a e^{i\langle e^{sA}x,h\rangle - \frac{1}{2}\langle Q_s h, h\rangle} ds - e^{i\langle x,h\rangle}\right| \leq 2.$$

Then $\mathcal{E}_A(H) \subset \overline{\mathcal{I}_A(H)}^\pi$. Consequently, in view of Proposition 3.1.3, $\mathcal{I}_A(H)$ is π-dense in $C_b(H)$. Now let $t > 0$. By taking into account (3.4), we can apply the Fubini theorem to find

$$R_t\left(\int_0^a e^{i\langle e^{sA}\cdot,h\rangle - \frac{1}{2}\langle Q_s h, h\rangle} ds\right)(x)$$

$$= \int_0^a e^{i\langle e^{(t+s)A}x,h\rangle - \frac{1}{2}\langle Q_t e^{sA^*}h, e^{sA^*}h\rangle - \frac{1}{2}\langle Q_s h, h\rangle} ds$$

$$= \int_0^a e^{i\langle e^{(t+s)A}x,h\rangle - \frac{1}{2}\langle Q_{t+s}h, h\rangle} ds$$

$$= \int_0^{a+t} e^{i\langle e^{sA}x,h\rangle - \frac{1}{2}\langle Q_s h, h\rangle} ds - \int_0^t e^{i\langle e^{sA}x,h\rangle - \frac{1}{2}\langle Q_s h, h\rangle} ds,$$

since $\langle Q_t e^{sA^*}h, e^{sA^*}h\rangle = \langle e^{sA}Q_t e^{sA^*}h, h\rangle = \langle Q_{t+s}h, h\rangle - \langle Q_s h, h\rangle$. Then we have $R_t(\mathcal{I}_A(H)) \subset \mathcal{I}_A(H)$. Now we prove that $\mathcal{I}_A(H) \subset D(L)$. Let

$$\varphi(x) = \int_0^a e^{i\langle e^{sA}x,h\rangle - \frac{1}{2}\langle Q_s h, h\rangle} ds. \tag{3.12}$$

By (3.1) we have that

$$R_t\varphi(x) - \varphi(x) = \int_a^{a+t} e^{i\langle e^{sA}x,h\rangle - \frac{1}{2}\langle Q_s h, h\rangle} ds - \int_0^t e^{i\langle e^{sA}x,h\rangle - \frac{1}{2}\langle Q_s h, h\rangle} ds.$$

This implies

$$\lim_{t \to 0^+} \frac{R_t \varphi(x) - \varphi(x)}{t} = e^{i\langle e^{aA}x, h \rangle - \frac{1}{2}\langle Q_a h, h \rangle} - e^{i\langle x, h \rangle} \qquad (3.13)$$

and

$$|R_t \varphi(x) - \varphi(x)| \le 2t.$$

Then $\varphi \in D(L)$ and by Proposition 2.2.6 follows that $\mathcal{I}_A(H)$ is a π-core for $(L, D(L))$. In order to prove (3.11), it is sufficient to take φ as in (3.12). By a straightforward computation we find that for any $x \in H$ it holds

$$\frac{1}{2}\mathrm{Tr}[QD^2\varphi(x)] + \langle x, A^*D\varphi(x) \rangle$$

$$= \int_0^a \left(i\langle A^* e^{sA^*}h, x \rangle - \frac{1}{2}\langle e^{sA}Qe^{sA^*}h, h \rangle \right) e^{i\langle e^{sA}x, h \rangle - \frac{1}{2}\langle Q_s h, h \rangle} ds$$

$$= \int_0^a \frac{\partial}{\partial s} e^{i\langle e^{sA}x, h \rangle - \frac{1}{2}\langle Q_s h, h \rangle} ds$$

$$= e^{i\langle e^{aA}x, h \rangle - \frac{1}{2}\langle Q_a h, h \rangle} - e^{i\langle x, h \rangle},$$

cf. Example 3.1.4. By taking into account (3.13), it follows that (3.11) holds. □

We are now able to prove the main result of this chapter

Theorem 3.1.6. *Let $(R_t)_{t \ge 0}$ be the Ornstein-Uhlenbeck semigroup (3.3) and let $L_0 : \mathcal{I}_A(H) \subset C_b(H) \to C_b(H)$ be the differential operator*

$$L_0\varphi(x) = \frac{1}{2}\mathrm{Tr}[QD^2\varphi(x)] + \langle x, A^*D\varphi(x) \rangle, \qquad \varphi \in \mathcal{I}_A(H)$$

Then, for any $\mu \in \mathcal{M}(H)$ there exists a unique family of measures $\{\mu_t, \; t \ge 0\} \subset \mathcal{M}(H)$ fulfilling (3.6) and the measure equation

$$\int_E \varphi(x)\mu_t(dx) - \int_H \varphi(x)\mu(dx) = \int_0^t \left(\int_H L_0\varphi(x)\mu_s(dx) \right) ds,$$

$$(3.14)$$

for any $\varphi \in \mathcal{I}_A(H)$, $t \ge 0$. Moreover, $\mu_t = R_t^*\mu$, $t \ge 0$.

Proof. By Proposition 3.1.5 we have that $\mathcal{I}_A(H)$ is a π-core for $(L, D(L))$ and that $L\varphi = L_0\varphi$, for any $\varphi \in \mathcal{I}_A(H)$. So it is easy to see that $R_t^*\mu$,

$t \geq 0$ is a solution of the measure equation (3.14). Hence, if $\varphi \in D(L)$ there exists a sequence[2] $(\varphi_n)_{n\in\mathbb{N}} \subset \mathcal{I}_A(H)$ such that

$$\lim_{n\to\infty} \varphi_n \stackrel{\pi}{=} \varphi, \quad \lim_{n\to\infty} L_0\varphi_n \stackrel{\pi}{=} K\varphi.$$

For any $t \geq 0$ we find

$$\int_H \varphi(x)\mu_t(dx) - \int_H \varphi(x)\mu(dx)$$

$$= \lim_{n\to\infty} \left(\int_H \varphi_n(x)\mu_t(dx) - \int_H \varphi_n(x)\mu(dx) \right)$$

$$= \lim_{n\to\infty} \int_0^t \left(\int_H K_0\varphi_n(x)\mu_s(dx) \right) ds.$$

Now observe that for any $s \geq 0$ it holds

$$\lim_{n\to\infty} \int_H L_0\varphi_n(x)\mu_s(dx) = \int_H L\varphi(x)\mu_s(dx)$$

and

$$\left| \int_H L_0\varphi_n(x)\mu_s(dx) \right| \leq \sup_{n\in\mathbb{N}} \|L_0\varphi_n\|_0 |\mu_s|_{TV}(H).$$

Hence, by taking into account (3.6) and that $\sup_{n\in\mathbb{N}} \|L_0\varphi_n\|_0 < \infty$, we can apply the dominated convergence theorem to obtain

$$\lim_{n\to\infty} \int_0^t \left(\int_H L_0\varphi_n(x)\mu_s(dx) \right) ds = \int_0^t \left(\int_H L\varphi(x)\mu_s(dx) \right) ds.$$

So, $\mu_t, t \geq 0$ is also a solution of the measure equation (3.6), (3.7). Since by Theorem 3.1.1 such a solution is unique, it follows that the measure equation (3.6), (3.14) has a unique solution, defined by $R_t^*\mu, t \geq 0$.   □

3.2. Absolute continuity with respect to the invariant measure

We consider the case when the semigroup R_t has a unique invariant measure μ. The aim of this section is to study the absolute continuity of the family $(\mu_t)_{t\geq0}$, solution of (3.6), (3.7), with $\mu_0 = \rho\mu$, where $\rho \in L^1(H, \mu)$.

We recall that a necessary and sufficient condition that guarantees existence of an invariant measure is that

$$\sup_{t\geq0} \mathrm{Tr}[Q_t] < \infty,$$

[2] For simplicity we assume that this sequence has only one index.

see [23], Theorem 11.7. For simplicity, in this section we shall assume that Hypothesis 1.2.1 holds and that $\omega < 0$. These conditions imply that the operator

$$Q_\infty = \int_0^\infty e^{tA} Q e^{tA^*} dt$$

is well defined and of trace class (see, for instance, [23]). We also have that $\mu = N_{Q_\infty}$ is the unique invariant measure for the semigroup $(R_t)_{t \geq 0}$ and that

$$\lim_{t \to +\infty} R_t \varphi(x) = \int_H \varphi(x) \mu(dx) = \langle \varphi, \mu \rangle,$$

for all $\varphi \in C_b(H)$, $x \in H$. The last statement means that the dynamical system $(H, \mathcal{B}(H), \mu, (R_t)_{t \geq 0})$ is *strongly mixing*.

For $p \geq 1$, we consider the functional space $L^p(H; \mu)$. For any $\varphi \in C_b(H)$ we have

$$\int_H |R_t \varphi(x)|^p \mu(dx) \leq \int_H R_t |\varphi|^p(x) \mu(dx) = \int_H |\varphi(x)|^p \mu(dx),$$

since μ is the invariant for R_t. This allows us to extend the Ornstein-Uhlenbeck semigroup $(R_t)_{t \geq 0}$ to a strongly continuous semigroup of contractions, still denoted it by $(R_t)_{t \geq 0}$, on $L^p(H; \mu)$. When $p = 2$, we shall denote the scalar product of the Hilber space $L^2(H; \mu)$ by

$$\langle \varphi, \psi \rangle_{L^2(H;\mu)}, \quad \varphi, \psi \in L^2(H; \mu).$$

3.3. The adjoint of R_t in $L^2(H; \mu)$

By Theorem 3.1.1, we have $\mu_t = R_t^* \mu_0$, that is

$$\int_H \varphi(x) \mu_t(dx) = \int_H R_t \varphi(x) \mu_0(dx) = \int_H R_t \varphi(x) \rho(x) \mu(dx),$$

for any $\varphi \in C_b(H)$, $t \geq 0$. Then, it is natural to study the adjoint of R_t in the space $L^2(H, \mu)$.

Following Chojnowska-Michalik and Goldys (see [10]), we shall give an explicit representation of the adjoint of R_t in the Hilbert space $L^2(H; \mu)$, by using the so called *second quantization operator*. We set

$$L_\mu(H) = \{T \in L(H) : \exists S \in L(H) \text{ such that } T = S Q_\infty^{1/2}\}.$$

It is easy to see that $T \in L_\mu(H)$ if and only if

$$T|_{Q_\infty^{1/2}(H)} \in \mathcal{L}\left((Q_\infty^{1/2}, H); H\right),$$

where $(Q_\infty^{1/2}, H)$ is the Banach space endowed with the norm $\|x\|_{Q_\infty^{1/2}(H)} = |Q_\infty^{1/2}x|$.

Consequently, since $Q_\infty^{1/2}(H)$ is dense in H, the space L_μ is dense in $L(H)$ with respect to the pointwise convergence.

Let us define a linear mapping

$$F : L_\mu(H) \to L^2(H, \mu; H)$$

by setting

$$F(T)x = Q_\infty Sx,$$

where $S \in L(H)$ is such that $T = SQ_\infty^{1/2}$. It is easy to see that

$$\int_H |Fx|^2 \mu(dx) = \mathrm{Tr}[Q_\infty^{1/2}TT^*Q_\infty^{1/2}].$$

Then F is extendible by density to all $L(H)$. We shall still denote this extension by F, and we shall write

$$F(T)x = Q_\infty^{1/2}TQ_\infty^{-1/2}x.$$

Clearly, F is not, in general, a bounded linear operator. Let us define, for any contraction $T \in \mathcal{L}(H)$, the linear operator

$$\Gamma : \{T \in \mathcal{L}(H) : \|T\|_{\mathcal{L}(H)} \le 1\} \to \mathcal{L}(L^p(H, \mu))$$

by setting

$$(\Gamma(T)\varphi)(x) = \int_H \varphi(Q_\infty^{1/2}T^*Q_\infty^{-1/2}x + Q_\infty^{1/2}\sqrt{I - T^*T}Q_\infty^{-1/2}y)\mu(dy),$$

for all $\varphi \in L^p(H; \mu)$. It is easy to check that $\Gamma(T)$ is still a contraction and that $(\Gamma(T))^* = \Gamma(T^*)$. The operator Γ is called the *second quantization operator*. For details we refer to [10, 24].

We have the next result, proved in [10].

Theorem 3.3.1. *Assume that for any $t > 0$ it holds*

$$e^{tA}(Q_\infty^{1/2}(H)) \subset Q_\infty^{1/2}(H). \tag{3.15}$$

Then, for all $t > 0$ and $\varphi \in L^2(H; \mu)$ we have

$$R_t^*\varphi(x) = \left(\Gamma(Q_\infty^{-1/2}e^{tA}Q_\infty^{1/2})\varphi\right)(x).$$

Remark 3.3.2. Condition (3.15) is weaker than to require that the OU semigroup R_t enjoys the *strong Feller property*. We say that a semigroup $(P_t)_{t \geq 0}$ on $B_b(H)$, the Banach space of all bounded and Borel functions $\varphi : H \to \mathbb{R}$, enjoys the strong Feller property when it holds

$$P_t \varphi \in C_b(H)$$

for any $t > 0$, $\forall \varphi \in B_b(H)$. This property is equivalent to require that for all $t > 0$ it holds

$$e^{tA}(H) \subset Q_t^{1/2}(H), \tag{3.16}$$

see [23], Theorem 9.19.

We recall that in general the adjoint of $(R_t)_{t \geq 0}$ in $L^2(H; \mu)$ is not an Ornstein-Uhlenbeck semigroup. However, the next proposition gives a sufficient condition in order to have this property.

Proposition 3.3.3. *Assume that $Q_\infty(H) \subset D(A^*)$ and that the operator*

$$A_1 x = Q_\infty A^* Q_\infty^{-1}, \; x \in D(A_1) = \{x \in Q_\infty(H) : Q_\infty^{-1} x \in D(A^*)\}$$

generates a C_0-semigroup given by $e^{tA_1} = Q_\infty e^{tA^*} Q_\infty^{-1}$. Then the adjoint of R_t in $L^2(H; \mu)$ is the operator R_t^* defined by

$$R_t^* \varphi(x) = \int_H \varphi(y) N_{e^{tA_1}, Q_{1,t}}(dy),$$

where

$$Q_{1,t} x = \int_0^t e^{sA_1} Q e^{sA_1^*} x \, ds, \quad x \in H, \; t \geq 0.$$

Proof. See [26], Proposition 10.1.9. □

3.3.1. Absolute continuity of μ_t

Theorem 3.3.4. *Let $\mu_0(dx) = \rho(x)\mu(dx)$, where $\rho \in L^1(H; \mu)$, and let the family of measure $\{\mu_t, \; t \geq 0\}$ be the solution of (3.6), (3.7). If (3.16) holds, then for all $t \geq 0$ the measure μ_t is absolutely continuous with respect to μ and it satisfies*

$$\mu_t(dx) - \left(\Gamma(Q_\infty^{-1/2} e^{tA} Q_\infty^{1/2}) \rho \right)(x)\mu(dx).$$

Proof. Let $\{\rho_n\}_{n \in \mathbb{N}}$ be a sequence in $L^2(H, \mu)$ that converges to ρ in $L^1(H; \mu)$. Since for all $t \geq 0$ we have $R_t(C_b(H)) \subset C_b(H) \subset L^2(H; \mu)$, it holds

$$\int_H |R_t \varphi(x)(\rho_n(x) - \rho(x))| \mu(dx) \leq \|\varphi\|_0 \int_H |\rho_n(x) - \rho(x)| \mu(dx), \tag{3.17}$$

for all $\varphi \in C_b(H)$, $n \in \mathbb{N}$. Clearly, this implies

$$\lim_{n\to\infty} \langle R_t\varphi, \rho_n \rangle_{L^2(H;\mu)} = \langle \varphi, \mu_t \rangle,$$

for all $\varphi \in C_b(H)$. Now let us set $S(t) = Q_\infty^{-1/2}e^{tA}Q_\infty^{1/2}$. By Theorem 3.3.1 we have

$$\langle R_t\psi_1, \psi_2 \rangle_{L^2(H;\mu)} = \langle \psi_1, \Gamma(S(t))\psi_2 \rangle_{L^2(H;\mu)}, \qquad (3.18)$$

for any $\psi_1, \psi_2 \in L^2(H; \mu)$. Since $\Gamma(S(t)) \in \mathcal{L}(L^1(H; \mu))$, by a computation as above we obtain

$$\lim_{n\to\infty} \langle \varphi, \Gamma(S(t))\rho_n \rangle_{L^2(H,\mu)} = \int_H \varphi(x)\,(\Gamma(S(t))\rho)\,(x)\mu(dx),$$

for any $\varphi \in C_b(H)$. Finally, taking into account (3.17), (3.18) for any $\varphi \in C_b(H)$ it follows

$$\begin{aligned}
\langle \varphi, \mu_t \rangle &= \langle R_t\varphi, \mu_0 \rangle \\
&= \lim_{n\to\infty} \langle R_t\varphi, \rho_n \rangle_{L^2(H;\mu)} = \lim_{n\to\infty} \langle \varphi, \Gamma(S(t))\rho_n \rangle_{L^2(H;\mu)} \\
&= \int_H \varphi(x)\,(\Gamma(S(t))\rho)\,(x)\mu(dx).
\end{aligned}$$

This concludes the proof. $\qquad\qquad\qquad\qquad\qquad\qquad\qquad\qquad$ $\square$

3.3.2. The case of a symmetric Ornstein-Uhlenbeck semigroup

A particular class of Ornstein-Uhlenbeck processes are the so called *reversible Ornstein-Uhlenbeck processes*, which arise in the theory of Interacting Particle System and other areas of Mathematical Physics. We are interested to find necessary and sufficient conditions on A and Q, in order to have

$$\langle R_t\varphi, \psi \rangle_{L^2(H;\mu)} = \langle \varphi, R_t\psi \rangle_{L^2(H;\mu)},$$

where μ is the invariant measure for R_t and $\varphi, \psi \in L^2(H; \mu)$. This problem was solved in [47] in the case $Q = I$. A characterization for general symmetric Ornstein-Uhlenbeck semigroups of the form (3.3) has been given by Chojnowska-Michalik and Goldys in [11] as follows

Theorem 3.3.5. *The following conditions are equivalent*

(i) *The semigroup* $(R_t)_{t\geq 0}$ *is symmetric in* $L^2(H; \mu)$;
(ii) *if* $x \in D(A^*)$ *then* $Qx \in D(A)$ *and* $AQx = QA^*x$;
(iii) $e^{tA}Q = Qe^{tA^*}$ *for all* $t \geq 0$.

See Theorem 2.7 in [11].

Corollary 3.3.6. *Let $(R_t)_{t\geq 0}$ be symmetric. Then the following holds*

(i) $Q_\infty(H) \subset D(A)$ *and the operator* $AQ_\infty = -\frac{1}{2}Q$ *is bounded, symmetric and negative;*

(ii) $Q(H) \subset A(H);$

(iii) *if* $\ker A = \{0\}$, *then*

$$Q_\infty = -\frac{1}{2}A^{-1}Q = -\frac{1}{2}\overline{Q(A^*)^{-1}}.$$

See Corollary 2.5 in [11].

Chapter 4
Bounded perturbations of OU operators

4.1. Introduction and main results

We consider here the stochastic differential equation in H

$$\begin{cases} dX(t) = (AX(t) + F(X(t)))dt + BdW(t), & t \geq 0 \\ X(0) = x \in H, \end{cases} \tag{4.1}$$

where A, B, W are as in Hypothesis 1.2.1 and

Hypothesis 4.1.1. $F : H \to H$ is Lipschitz continuous and bounded.

Under Hypothesis 1.2.1, 4.1.1 equation (4.1) has a unique mild solution

$$X(t, x) = e^{tA}x + \int_0^t e^{(t-s)A} BdW(s) + \int_0^t e^{(t-s)A} F(X(s, x))ds, \tag{4.2}$$

(see, for instance, [23]). The transition semigroup $(P_t)_{t \geq 0}$ in $C_b(H)$ associated to equation (4.1) is defined by setting

$$P_t\varphi(x) = \mathbb{E}\big[\varphi(X(t, x))\big], \quad \varphi \in C_b(H), \ t \geq 0, \ x \in H. \tag{4.3}$$

Since $X(t, x)$ is continuous in mean square, as easily checked the semigroup $(P_t)_{t \geq 0}$ is a stochastically continuous Markov semigroup (cf. Proposition 4.2.1). This allows us to define the infinitesimal generator $(K, D(K))$ of $(P_t)_{t \geq 0}$ as in (6), by setting

$$\begin{cases} D(K) = \left\{ \varphi \in C_b(H) : \exists g \in C_b(H), \lim_{t \to 0^+} \dfrac{P_t\varphi(x) - \varphi(x)}{t} = g(x), \right. \\ \qquad\qquad \left. \forall x \in H, \ \sup_{t \in (0,1)} \left\| \dfrac{P_t\varphi - \varphi}{t} \right\|_0 < \infty \right\} \\ \\ K\varphi(x) = \lim_{t \to 0^+} \dfrac{P_t\varphi(x) - \varphi(x)}{t}, \quad \varphi \in D(K), \ x \in H. \end{cases} \tag{4.4}$$

4.2. The transition semigroup and its infinitesimal generator

We begin with showing that the transition semigroup $(P_t)_{t\geq 0}$ in (4.3) is a stochastically continuous Markov semigroup in $C_b(H)$.

Proposition 4.2.1. *Under Hypothesis 1.2.1, 4.1.1 the transition semigroup $(P_t)_{t\geq 0}$ defined in (4.3) is a stochastically continuous Markov semigroup in $C_b(H)$.*

Proof. The proof of the fact that $(P_t)_{t\geq 0}$ maps $C_b(H)$ into $C_b(H)$ and that it is a semigroup of operators may be found in [13, Proposition 3.9]. We also have $P_t\varphi(x) = \int_H \varphi(y)\pi_t(x, dy)$, where $\pi_t(x, \cdot)$ is the probability Borel measure on H defined by $\pi_t(x, \Gamma) = \mathbb{P}(X(t, x) \in \Gamma), \forall \Gamma \in \mathcal{B}(H)$. Hence, the semigroup $(P_t)_{t\geq 0}$ is Markovian. Finally, since $X(t, x)$ fulfills (4.2), it follows easily that for any $\varphi \in C_b(H)$, $x \in H$ the function $H \to \mathbb{R}, t \to P_t\varphi(x)$ is continuous. $\square$

4.2.1. Comparison with the OU operator

According to Chapter 3, we consider the OU semigroup $(R_t)_{t\geq 0}$ under Hypothesis 1.2.1 given by formula (3.3) and its infinitesimal generator $(L, D(L))$, given by (3.5).

Proposition 4.2.2. *Under Hypothesis 1.2.1, 4.1.1 let $(L, D(L))$ be the infinite-simal generator of the OU semigroup $(R_t)_{t\geq 0}$, and let $(K, D(K))$ be the infinitesimal generator of the semigroup $(P_t)_{t\geq 0}$ in $C_b(H)$.*
Then $D(K) \cap C_b^1(H) = D(L) \cap C_b^1(H)$ and for any $\varphi \in D(L) \cap C_b^1(H)$ we have $K\varphi = L\varphi + \langle D\varphi, F\rangle$.

Proof. Let $X(t, x)$ be the solution of equation (4.2) and let us set

$$Z_A(t, x) = e^{tA}x + \int_0^t e^{(t-s)A} Q^{1/2} dW(s).$$

Let $\varphi \in D(L) \cap C_b^1(H)$. By taking into account that

$$X(t, x) = Z_A(t, x) + \int_0^t e^{(t-s)A} F(X(t, x)) ds,$$

by the Taylor formula we have that $\mathbb{P}$-a.s. it holds

$$\varphi(Z_A(t, x)) = \varphi(Z_A(t, x)) - \varphi(X(t, x)) + \varphi(X(t, x)) = \varphi(X(t, x))$$
$$- \int_0^1 \left\langle D\varphi(\xi Z_A(t, x) + (1 - \xi)X(t, x)), \int_0^t e^{(t-s)A} F(X(t, x)) ds \right\rangle d\xi.$$

Then we have

$$R_t\varphi(x) - \varphi(x) = \mathbb{E}\big[\varphi(Z_A(t, x))\big] - \varphi(x) = P_t\varphi(x) - \varphi(x)$$

$$- \mathbb{E}\left[\int_0^1 \Big\langle D\varphi(\xi Z_A(t, x) + (1-\xi)X(t, x)), \int_0^t e^{(t-s)A} F(X(t, x))ds \Big\rangle d\xi\right].$$

Since $\varphi \subset D(L) \cap C_b^1(H)$, it follows easily that for any $x \in H$

$$\lim_{t \to 0^+} \frac{P_t\varphi(x) - \varphi(x)}{t} = L\varphi(x) + \langle D\varphi(x), F(x)\rangle$$

and

$$\sup_{t \in (0,1]} \left\|\frac{P_t\varphi - \varphi}{t}\right\|_0 \leq \sup_{t \in (0,1]} \left\|\frac{R_t\varphi - \varphi}{t}\right\|_0 + \|D\varphi\|_{C_b(H;\mathcal{L}(H))}\|F\|_{C_b(H;H)}$$
$$< \infty,$$

that implies $\varphi \in D(K)$ and $K\varphi = L\varphi + \langle D\varphi, F\rangle$. The opposite inclusion follows by interchanging the role of R_t and P_t in the Taylor formula. $\quad\square$

4.2.2. The Kolmogorov operator

We consider the Kolmogorov operator associated to equation 4.1

$$K_0\varphi(x) = \frac{1}{2}\text{Tr}\big[QD^2\varphi(x)\big] + \langle x, A^*D\varphi(x)\rangle + \langle D\varphi(x), F(x)\rangle, \quad (4.5)$$

where $x \in H$, $\varphi \in \mathcal{I}_A(H)$ (the space $\mathcal{I}_A(H)$ has been introduced in the previous chapter).

Theorem 4.2.3. *The operator $(K, D(K))$ is an extension of K_0, and for any $\varphi \in \mathcal{I}_A(H)$ we have $K\varphi = K_0\varphi$.*

Proof. Note that $\mathcal{I}_A(H) \subset C_b^1(H)$. Since by Proposition 3.1.5 we have $\mathcal{I}_A(H) \subset D(L)$, by Proposition 4.2.2 we have $\mathcal{I}_A(H) \subset D(K)$ and $K\varphi = L\varphi + \langle D\varphi, F\rangle$, for any $\psi \in \mathcal{I}_A(H)$. Finally, by taking into account (3.11), it follows that $K\varphi = K_0\varphi$ holds for any $\varphi \in \mathcal{I}_A(H)$. $\quad\square$

4.3. A π-core for $(K, D(K))$

We now prove that $\mathcal{I}_A(H)$ is a π-core for K. We need the following approximation result

Lemma 4.3.1. *Under the hypothesis of Proposition* 4.2.2, *let* $\varphi \in D(L) \cap C_b^1(H)$. *Then there exists* $m \in \mathbb{N}$ *and an* m-*indexed sequence* $(\varphi_{n_1,\dots,n_m}) \subset \mathcal{I}_A(H)$ *such that*

$$\lim_{n_1 \to \infty} \cdots \lim_{n_m \to \infty} \varphi_{n_1,\dots,n_m} \overset{\pi}{=} \varphi, \qquad (4.6)$$

$$\lim_{n_1 \to \infty} \cdots \lim_{n_m \to \infty} \frac{1}{2} \text{Tr}\left[QD^2\varphi_{n_1,\dots,n_m}\right] + \langle \cdot, A^*D\varphi_{n_1,\dots,n_m} \rangle \overset{\pi}{=} L\varphi, \qquad (4.7)$$

and for any $h \in H$

$$\lim_{n_1 \to \infty} \cdots \lim_{n_m \to \infty} \langle D\varphi_{n_1,\dots,n_m}, h \rangle \overset{\pi}{=} \langle D\varphi, h \rangle. \qquad (4.8)$$

Proof. We observe that the results of Proposition 3.1.3 also holds by approximations with functions in $\mathcal{I}_A(H)$. Indeed, let $(\varphi_{n_1,n_2,n_3}) \subset \mathcal{E}_A(H)$ as in Proposition 3.1.3. By setting, for any $n_1, n_2, n_3, n_4 \in \mathbb{N}$

$$\varphi_{n_1,n_2,n_3,n_4}(x) = n_4 \int_0^{\frac{1}{n_4}} R_t \varphi_{n_1,n_2,n_3}(x)dt$$

we have, according to (3.4), that $\varphi_{n_1,n_2,n_3,n_4} \in \mathcal{I}_A(H)$. Clearly,

$$\lim_{n_1 \to \infty} \cdots \lim_{n_4 \to \infty} \varphi_{n_1,n_2,n_3,n_4} \overset{\pi}{=} \varphi.$$

Moreover, since $D(R_t f) = e^{tA^*} R_t(D\varphi)$ (*cf.*, *e.g.*, [26], Proposition 6.2.9), we find that for any $h \in H$ it holds

$$\langle D\varphi_{n_1,n_2,n_3,n_4}(x), h \rangle = n_4 \int_0^{\frac{1}{n_4}} R_t\big(\langle D\varphi_{n_1,n_2,n_3}(\cdot), e^{tA}h \rangle\big)(x)dt.$$

Hence,

$$\lim_{n_1 \to \infty} \cdots \lim_{n_4 \to \infty} \langle D\varphi_{n_1,n_2,n_3,n_4}, h \rangle \overset{\pi}{=} \langle D\varphi, h \rangle.$$

We now construct the desired approximation for $\varphi \in D(L) \cap C_b^1(H)$. Let $\varphi \in D(L) \cap C_b^1(H)$ and $(\varphi_{n_2}) \subset \mathcal{I}_A(H)$ as above (we denote this sequence with one index to avoid heavy notations). By setting (φ_{n_1,n_2,n_3}) as in (2.5) with R_t instead of P_t, we have that (4.6), (4.7) hold, by the same argument of the proof of Proposition 2.2.6.

We now observe that for any $n_1, n_2, n_3 \in \mathbb{N}$, the function φ_{n_1,n_2,n_3} is differentiable in every $x \in H$ along any direction $h \in H$, with differential

$$\langle D\varphi_{n_1,n_2,n_3}(x), h \rangle = \varphi_{n_1,n_2,n_3}(x) = \frac{1}{n_3} \sum_{i=1}^{n_3} R_{\frac{i}{n_1 n_3}}\big(\langle D\varphi_{n_2}(\cdot), e^{\frac{i}{n_1 n_3}A}h \rangle\big)(x).$$

Moreover,

$$\sup_{n_1,n_2,n_3\in\mathbb{N}} \|\langle D\varphi_{n_1,n_2,n_3}, h\rangle\|_0 \leq \sup_{n_2} \|D\varphi_{n_2}\|_{C_b(H;H)} \sup_{0\leq t\leq 1} \|e^{tA}\|_{\mathcal{L}(H)} |h|$$

$$< \infty.$$

Now by arguing as for Proposition 2.2.6, yields (4.8). □

4.3.1. The case $F \in C_b^2(H; H)$

The following proposition is proved in [13], section 3.3.

Proposition 4.3.2. *Let us assume Hypothesis 1.2.1, 4.1.1 and that $F \in C_b^2(H; H)$, that is $F : H \to H$ is twice differentiable with bounded differentials. Then the semigroup $(P_t)_{t\geq 0}$ defined in (4.3) maps $C_b^1(H)$ into $C_b^1(H)$, and for any $f \in C_b^1(H)$, $h \in H$ we have*

$$\langle DP_t f(x), h\rangle = \mathbb{E}\big[\langle Df(X(t, x)), \eta^h(t, x)\rangle\big],$$

where $\eta^h(t, x)$ is the mild solution of the differential equation with random coefficients in H

$$\begin{cases} \dfrac{d}{dt}\eta^h(t, x) = A\eta^h(t, x) + \langle DF(X(t, x)), \eta^h(t, x)\rangle, & t > 0, \\ \eta^h(0, x) = h. \end{cases}$$

Corollary 4.3.3. *Under the hypothesis of Proposition 4.3.2, let $(K, D(K))$ be the infinitesimal generator of $(P_t)_{t\geq 0}$. Then, for any $\lambda > 0, \omega + M\|DF\|_0$, the resolvent $R(\lambda, K)$ of K at λ maps $C_b^1(H)$ into $C_b^1(H)$ and it holds*

$$\|DR(\lambda, K)f\|_{C_b(H;H)} \leq \frac{M\|Df\|_{C_b(H;H)}}{\lambda - (\omega + M\|DF\|_{C_b(H;\mathcal{L}(H))})}, \quad f \in C_b^1(H). \quad (4.9)$$

Proof. Let $f \in C_b^1(H)$. For any $t \geq 0$, $P_t f \in C_b^1(H)$ and for any $x, h \in H$ it holds

$$\langle DP_t f(x), h\rangle = \mathbb{E}\big[\langle Df(X(t, x)), \eta^h(t, x)\rangle\big],$$

where $\eta^h(t, x)$ is as in Proposition 4.3.2. It is also easy to see that[1]

$$|\eta^h(t, x)| \leq Me^{(\omega + M\|DF\|)t}|h|,$$

---

[1] In order to avoid heavy notations we set $\|DF\| = \|DF\|_{C_b(H;\mathcal{L}(H))}$.

see, *e.g.*, [13], Theorem 3.6. Hence, by (vi) of Theorem 2.2.4, we have

$$|\langle DR(\lambda, K)f(x), h\rangle| = \left|\int_0^\infty e^{-\lambda t}\mathbb{E}\big[\langle Df(X(t, x)), \eta^h(t, x)\rangle\big]dt\right|$$

$$\leq M\|Df\|_{C_b(H;H)} \int_0^\infty e^{-\lambda t}e^{(\omega+M\|DF\|)t}|h|dt$$

$$= \frac{M\|Df\|_{C_b(H;H)}}{\lambda - (\omega + M\|DF\|_{C_b(H;\mathcal{L}(H))})}|h|,$$

for any $h \in H$. Therefore, (4.9) follows.   □

Proposition 4.3.4. *Let us assume that Hypothesis 1.2.1, 4.1.1 hold and let $F \in C_b^2(H; H)$. Denote by $(P_t)_{t\geq 0}$ the transition semigroup defined in (4.3), let $(K, D(K))$ be its infinitesimal generator. Then, the set $\mathcal{I}_A(H)$ is a π-core for $(K, D(K))$, and for any $\varphi \in D(K)$ there exists $m \in \mathbb{N}$ and an m-indexed sequence $(\varphi_{n_1,...,n_m}) \subset \mathcal{I}_A(H)$ such that*

$$\lim_{n_1\to\infty} \cdots \lim_{n_m\to\infty} K_0\varphi_{n_1,...,n_m} \overset{\pi}{=} K\varphi. \tag{4.10}$$

Proof. Let $\varphi \in D(L) \cap C_b^1(H)$. By Proposition 4.2.2 we have that $\varphi \in D(K) \cap C_b^1(H)$. Hence, by (i) of Theorem 2.2.4 we have $P_t\varphi \in D(K)$ and by Proposition 4.3.2 we have $P_t\varphi \in C_b^1(H)$, for any $t \geq 0$. So $P_t\colon D(L)\cap C_b^1(H) \to D(L)\cap C_b^1(H)$, for any $t \geq 0$. Moreover, $\mathcal{I}_A(H) \subset D(L)\cap C_b^1(H)$ and so $D(L)\cap C_b^1(H)$ is π-dense in $C_b(H)$, in view of the fact that $\mathcal{I}_A(H)$ is π-dense in $C_b(H)$ (*cf.* Proposition 3.1.5). Therefore, by Proposition 2.2.6, $D(L)\cap C_b^1(H)$ is a π-core for $(K, D(K))$. So there exists a sequence $(\varphi_m) \subset \mathcal{I}_A(H)$ (we denote this sequence with one index to avoid heavy notations) such that $L\varphi_m + \langle D\varphi_m, F\rangle \overset{\pi}{\to} K\varphi$, as $m \to \infty$. Now, thanks to Lemma 4.3.1, we can approximate any φ_m by a sequence $(\varphi_{m,n}) \subset \mathcal{I}_A(H)$ in such a way that $\varphi_{m,n} \overset{\pi}{\to} \varphi_m$, $L\varphi_{m,n} \overset{\pi}{\to} L\varphi_m$ as $n \to \infty$ and $\langle D\varphi_{m,n}, h\rangle \overset{\pi}{\to} \langle D\varphi_m, h\rangle$ as $n \to \infty$, for any $h \in H$. Since $F : H \to H$ is bounded, we have $\langle D\varphi_{m,n}, F\rangle \overset{\pi}{\to} \langle D\varphi_m, F\rangle$ as $n \to \infty$. Finally, since $\varphi_{m,n} \in \mathcal{I}_A(H)$ by Theorem 4.2.3 it follows (4.10).   □

4.3.2. The case when F is Lipschitz

Theorem 4.2.3 shows that K is an extension of K_0, and that $K\varphi = K_0\varphi$, $\forall\varphi \in \mathcal{I}_A(H)$.

We now show that $\mathcal{I}_A(H)$ is a π-core for K.

Theorem 4.3.5. *Under the Hypothesis of Theorem 4.2.3, the set $\mathcal{I}_A(H)$ is a π-core for $(K, D(K))$.*

Proof. We denote by L_F the Lipschitz constant of F. Let $\varphi \in D(K)$, $\lambda > \max\{0, \omega + L_F\}$ and set $f = \lambda\varphi - K\varphi$. Since $C_b^1(H)$ is dense in $C_b(H)$ with respect to the supremum norm (see [34]), there exists a sequence $(f_{n_1}) \subset C_b^1(H)$ such that $\|f_{n_1} - f\|_0 \to 0$ as $n_1 \to \infty$. Clearly, if $\varphi_{n_1} = R(\lambda, K)f_{n_1}$ we have

$$\lim_{n_1\to\infty} K\varphi_{n_1} \overset{\pi}{=} K\varphi. \tag{4.11}$$

Now we consider a sequence of functions $(F_{n_2})_{n_2\in\mathbb{N}} \subset C_b^2(H; H)$ such that

$$\lim_{n_2\to\infty} F_{n_2}(x) = F(x), \quad \forall x \in H \tag{4.12}$$

and

$$\sup_{n_2\in\mathbb{N}} \|F_{n_2}\|_{C_b(H;H)} \leq \|F\|_{C_b(H;H)}, \quad \sup_{n_2\in\mathbb{N}} \|DF_{n_2}\|_{C_b(H;\mathcal{L}(H))} \leq L_F. \tag{4.13}$$

This construction is not too difficult but technical and an example can be found in [13, Section 3.3.1]. Let $X^{n_2}(t, x)$ be the solution of (4.2) with F_{n_2} instead of F. It is straightforward to see that for any $T > 0, x \in H$

$$\lim_{n_2\to\infty} \sup_{t\in[0,T]} \mathbb{E}\left[|X^{n_2}(t, x) - X(t, x)|^2\right] = 0.$$

Hence, if $P_t^{n_2}$ is the transition semigroup associated to $X^{n_2}(t, x)$, we have that for any $\varphi \in C_b(H)$

$$\lim_{n_2\to\infty} P_t^{n_2}\varphi \overset{\pi}{=} P_t\varphi.$$

We denote by $(K_{n_2}, D(K_{n_2}))$ the infinitesimal generator of the transition semigroup $\{P_t^{n_2}\}_{t\geq0}$, as in (2.2). We also set

$$K_{0,n_2}\varphi(x) = K_0\varphi(x) + \langle D\varphi(x), F_{n_2} - F(x)\rangle, \quad \varphi \in \mathcal{I}_A(H), \; x \in H.$$

If $R(\lambda, K_{n_2})$ is the resolvent of K_{n_2} at λ (*cf.* (vi) of Theorem 2.2.4), we have

$$\lim_{n_2\to\infty} R(\lambda, K_{n_2})f \overset{\pi}{=} R(\lambda, K)f,$$

for any $f \in C_b(H)$. Setting $\varphi_{n_1,n_2} = R(\lambda, K_{n_2})f_{n_1}$, for any $n_1 \in \mathbb{N}$ we have

$$\lim_{n_2\to\infty} \varphi_{n_1,n_2} \overset{\pi}{=} \varphi_{n_1}, \quad \lim_{n_2\to\infty} K_{n_2}\varphi_{n_1,n_2} \overset{\pi}{=} K\varphi_{n_1}. \tag{4.14}$$

Moreover, since $F_{n_2} \in C_b^2(H; H)$, by Corollary 4.3.3 we have that $R(\lambda, K_{n_2}) : C_b^1(H) \to C_b^1(H)$ and

$$\|D\varphi_{n_1,n_2}\|_{C_b(H;H)} \leq \frac{M\|D\varphi_{n_1}\|_{C_b(H;H)}}{\lambda - (\omega + \|DF_{n_2}\|_{C_b(H;\mathcal{L}(H))})} \leq \frac{M\|D\varphi_{n_1}\|_{C_b(H;H)}}{\lambda - (\omega + L_F)},$$

for any $n_1, n_2 \in \mathbb{N}$. Consequently, by (4.12), (4.13) it follows

$$\lim_{n_2 \to \infty} \langle D\varphi_{n_1,n_2}, F - F_{n_2} \rangle \overset{\pi}{=} 0. \tag{4.15}$$

Since $f_{n_1} \in C_b^1(H)$, by Corollary 4.3.3 we have $\varphi_{n_1,n_2} \in D(K_{n_2}) \cap C_b^1(H)$. By Proposition 4.3.4, for any $n_1, n_2 \in \mathbb{N}$ we can find a sequence $(\varphi_{n_1,n_2,n_3}) \subset \mathcal{I}_A(H)$ such that

$$\lim_{n_3 \to \infty} K_{0,n_2}\varphi_{n_1,n_2,n_3} \overset{\pi}{=} L\varphi_{n_1,n_2} + \langle D\varphi_{n_1,n_2}, F_{n_2} \rangle = K_{n_2}\varphi_{n_1,n_2}. \tag{4.16}$$

Hence we have

$$K_0\varphi_{n_1,n_2,n_3} = K_{0,n_2}\varphi_{n_1,n_2,n_3} + \langle D\varphi_{n_1,n_2,n_3}, F - F_{n_2} \rangle$$

and by (4.14), (4.15), (4.16) it follows

$$\lim_{n_2 \to \infty} \lim_{n_3 \to \infty} K_0\varphi_{n_1,n_2,n_3}$$

$$\overset{\pi}{=} \lim_{n_2 \to \infty} K_{n_2}\varphi_{n_1,n_2} + \langle D\varphi_{n_1,n_2}, F - F_{n_2} \rangle \overset{\pi}{=} K\varphi_{n_1}.$$

Now the result follows by (4.11). □

4.4. The measure equation for K_0

The following result follows by Theorem 4.2.3, Theorem 4.3.5 and may be proved in essentially the same way as for Theorem 3.1.6.

Theorem 4.4.1. *Let $(P_t)_{t \geq 0}$ be the transition semigroup defined in (4.3) and let $(K, D(K))$ its infinitesimal generator. Then, for any $\mu \in \mathcal{M}(H)$ there exists a unique family of measures $\{\mu_t, \ t \geq 0\} \subset \mathcal{M}(H)$ fulfilling*

$$\int_0^T |\mu_t|_{TV}(H)dt < \infty, \quad T > 0; \tag{4.17}$$

and the measure equation

$$\int_E \varphi(x)\mu_t(dx) - \int_H \varphi(x)\mu(dx) = \int_0^t \left(\int_H K_0\varphi(x)\mu_s(dx) \right)ds, \tag{4.18}$$

for any $\varphi \in \mathcal{I}_A(H)$, $t \geq 0$, and the solution is given by $P_t^*\mu$, $t \geq 0$.

Chapter 5
Lipschitz perturbations
of Ornstein-Uhlenbeck operators

We consider here Kolmogorov operators with a Lispchitz continuous nonlinearity in the space $C_{b,1}(H)$.

The main novelties are discussed in Theorems 5.1.3, 5.1.4 and are contained in the submitted paper [38].

5.1. Introduction

Let us consider the stochastic differential equation in the Hilbert space H

$$\begin{cases} dX(t) = \big(AX(t) + F(X(t))\big)dt + B\,dW(t), & t \geq 0 \\ X(0) = x \in H, \end{cases} \tag{5.1}$$

where, behind Hypothesis 1.2.1 we assume that

Hypothesis 5.1.1. $F : H \to H$ is a Lipschitz continuous map. We set

$$\kappa = \sup_{\substack{x,y \in H \\ x \neq y}} \frac{|F(x) - F(y)|}{|x - y|}.$$

It is well known that under hypothesis 1.2.1 and (5.1.1) for any $x \in H$ problem (5.1) has a unique mild solution, that is a solution of the following integral equation

$$X(t,x) = e^{tA}x + \int_0^t e^{(t-s)A} B\,dW(s) + \int_0^t e^{(t-s)A} F(X(s,x))ds \tag{5.2}$$

for any $t \geq 0$. Moreover, a straightforward computation shows that for any $T > 0$ there exists $c > 0$ such that

$$\sup_{t \in [0,T]} |X(t,x) - X(t,y)| \leq c|x - y|, \quad \forall x, y \in H, \tag{5.3}$$

and

$$\sup_{t \in [0,T]} \mathbb{E}\big[|X(t,x)|\big] \leq c(1 + |x|), \quad x \in H, \tag{5.4}$$

where the expectation is taken with respect to $\mathbb{P}$. As we shall see in Proposition 5.1.6, estimates (5.3), (5.4) allow us to define the transition operator associated to equation (5.2) in the space $C_{b,1}(H)$, by the formula

$$P_t\varphi(x) = \mathbb{E}[\varphi(X(t,x))], \quad \varphi \in C_{b,1}(H), \ t \geq 0, \ x \in H. \tag{5.5}$$

Still by Proposition 5.1.6, we see that the family of operators $(P_t)_{t\geq0}$ maps $C_{b,1}(H)$ into $C_{b,1}(H)$ and enjoys the semigroup property, but it is not a strongly continuous semigroup. However, we can define the infinitesimal generator of $(P_t)_{t\geq0}$ in $C_{b,1}(H)$ in the following way

$$
\begin{cases}
D(K, C_{b,1}(H)) = \left\{ \varphi \in C_{b,1}(H) : \exists g \in C_{b,1}(H), \ \lim_{t\to0^+} \dfrac{P_t\varphi(x)-\varphi(x)}{t} = \right. \\
\left. = g(x), \ x \in H, \ \sup_{t\in(0,1)} \left\| \dfrac{P_t\varphi-\varphi}{t} \right\|_{0,1} < \infty \right\} \\
K\varphi(x) = \lim_{t\to0^+} \dfrac{P_t\varphi(x)-\varphi(x)}{t}, \quad \varphi \in D(K, C_{b,1}(H)), \ x \in H.
\end{cases}
\tag{5.6}
$$

The first result of the chapter is the following generalization of Theorems 2.2.3, 2.3.1

Theorem 5.1.2. *Let $(P_t)_{t\geq0}$ be the semigroup defined by (5.5) and let $(K, D(K, C_{b,1}(H)))$ be its infinitesimal generator in $C_{b,1}(H)$, defined by (5.6). Then, the formula*

$$\langle\varphi, P_t^*F\rangle_{\mathcal{L}(C_{b,1}(H),(C_{b,1}(H))^*)} = \langle P_t\varphi, F\rangle_{\mathcal{L}(C_{b,1}(H),(C_{b,1}(H))^*)}$$

defines a semigroup $(P_t^*)_{t\geq0}$ of linear and continuous operators on $(C_{b,1}(H))^*$ which maps $\mathcal{M}_1(H)$ into $\mathcal{M}_1(H)$. Moreover, for any $\mu \in \mathcal{M}_1(H)$ there exists a unique family of measures $\{\mu_t, \ t \geq 0\} \subset \mathcal{M}_1(H)$ such that

$$\int_0^T \left(\int_H |x||\mu_t|_{TV}(dx) \right) dt < \infty, \quad \forall T > 0 \tag{5.7}$$

and

$$\int_H \varphi(x)\mu_t(dx) - \int_H \varphi(x)\mu(dx) = \int_0^t \left(\int_H K\varphi(x)\mu_s(dx) \right) ds \tag{5.8}$$

for any $t \geq 0$, $\varphi \in D(K, C_{b,1}(H))$. Finally, the solution of (5.8) is given by $P_t^*\mu, \ t \geq 0$.

A natural question is to study the above problem replacing K with the *Kolmogorov* differential operator

$$K_0\varphi(x) = \frac{1}{2}\mathrm{Tr}\big[BB^*D^2\varphi(x)\big] + \langle x, A^*D\varphi(x)\rangle + \langle D\varphi(x), F(x)\rangle, \quad x \in H.$$
(5.9)

We stress the fact that the operator K is defined in an abstract way, whereas K_0 is a *concret* differential operator.

In order to study problem (5.8) with K_0 replacing K, we shall extend the notion of π-convergence in the spaces $C_{b,k}(H)$ and the related notion of π-core. We recall that the π-convergence has been introduced in Definition 2.1.1.

We have the following

Theorem 5.1.3. *Under Hypothesis* 1.2.1 *and* (5.1.1), *the operator* $(K, D(K, C_{b,1}(H)))$ *is an extension of* K_0, *and for any* $\varphi \in \mathcal{E}_A(H)$ *we have* $\varphi \in D(K, C_{b,1}(H))$ *and* $K\varphi = K_0\varphi$. *Finally,* $\mathcal{E}_A(H)$ *is a* π-core *for* $(K, D(K, C_{b,1}(H)))$.

As consequence we have the third main result of this chapter

Theorem 5.1.4. *For any* $\mu \in \mathcal{M}_1(H)$ *there exists an unique family of measures* $\{\mu_t, \ t \geq 0\} \subset \mathcal{M}_1(H)$ *fulfilling* (5.7) *and the measure equation*

$$\int_H \varphi(x)\mu_t(dx) - \int_H \varphi(x)\mu(dx) = \int_0^t \left(\int_H K_0\varphi(x)\mu_s(dx)\right) ds,$$
(5.10)

$t \geq 0$, $\varphi \in \mathcal{E}_A(H)$. *Moreover, the solution is given by* $P_t^*\mu$, $t \geq 0$.

Remark 5.1.5. *We shall work with* uniformly continuous *functions for convenience only. It is worth noticing that we can state all the result of this chapter (and also of other chapters) in spaces of* continuous *functions.*

5.1.1. The transition semigroup in $C_{b,1}(H)$

This section is devoted to studying the semigroup $(P_t)_{t\geq 0}$ in the space $C_{b,1}(H)$.

Proposition 5.1.6. *Formula* (5.5) *defines a semigroup of operators* $(P_t)_{t\geq 0}$ *in* $C_{b,1}(H)$, *and there exist a family of probability measures* $\{\pi_t(x, \cdot), \ t \geq 0, \ x \in H\} \subset \mathcal{M}_1(H)$ *and two constants* $c_0 > 0$, $\omega_0 \in \mathbb{R}$ *such that*

(i) $P_t \in \mathcal{L}(C_{b,1}(H))$ *and* $\|P_t\|_{\mathcal{L}(C_{b,1}(H))} \leq c_0 e^{\omega_0 t}$;

(ii) $P_t\varphi(x) = \int_H \varphi(y)\pi_t(x, dy)$, *for any* $t \geq 0$, $\varphi \in C_{b,1}(H)$, $x \in H$;

(iii) *for any $\varphi \in C_{b,1}(H)$, $x \in H$, the function $\mathbb{R}^+ \to \mathbb{R}$, $t \mapsto P_t\varphi(x)$ is continuous.*

(iv) *$P_t P_s = P_{t+s}$, for any $t, s \geq 0$ and $P_0 = I$;*

(v) *for any $\varphi \in C_{b,1}(H)$ and any sequence $(\varphi_n)_{n\in\mathbb{N}} \subset C_{b,1}(H)$ such that*

$$\lim_{n\to\infty} \frac{\varphi_n}{1+|\cdot|} \overset{\pi}{=} \frac{\varphi}{1+|\cdot|}$$

we have, for any $t \geq 0$,

$$\lim_{n\to\infty} \frac{P_t\varphi_n}{1+|\cdot|} \overset{\pi}{=} \frac{P_t\varphi}{1+|\cdot|}.$$

Proof. (i). Take $\varphi \in C_{b,1}(H)$, $t \geq 0$. We have to show that $P_t\varphi \in C_{b,1}(H)$, that is the function $x \mapsto (1+|x|)^{-1} P_t\varphi(x)$ is uniformly continuous and bounded. Take $\varepsilon > 0$ and let $\theta_\varphi : \mathbb{R}^+ \to \mathbb{R}$ be the modulus of continuity of $(1+|\cdot|)^{-1}\varphi$. We have

$$\frac{P_t\varphi(x)}{1+|x|} - \frac{P_t\varphi(y)}{1+|y|} = I_1(t, x, y) + I_2(t, x, y) + I_3(t, x, y),$$

where

$$I_1(t, x, y) = \mathbb{E}\left[\left(\frac{\varphi(X(t, x))}{1+|X(t, x)|} - \frac{\varphi(X(t, y))}{1+|X(t, y)|}\right)\frac{1+|X(t, x)|}{1+|x|}\right],$$

$$I_2(t, x, y) = \mathbb{E}\left[\frac{\varphi(X(t, y))}{1+|X(t, y)|}\left(\frac{|X(t, x)| - |X(t, y)|}{1+|x|}\right)\right],$$

$$I_3(t, x, y) = \mathbb{E}\left[\frac{\varphi(X(t, y))(1+|X(t, x)|)}{1+|X(t, y)|}\left(\frac{1}{1+|x|} - \frac{1}{1+|y|}\right)\right].$$

For $I_1(t, x, y)$ we have, by taking into account (5.3), (5.4), that there exists $c > 0$ such that

$$|I_1(t, x, y)| \leq \mathbb{E}\left[\theta_\varphi(|X(t, x) - X(t, y)|)\frac{1+|X(t, x)|}{1+|x|}\right]$$

$$\leq \theta_\varphi(c|x - y|)\frac{\mathbb{E}[1+|X(t, x)|]}{1+|x|} \leq c\theta_\varphi(c|x - y|).$$

Then, there exists $\delta_1 > 0$ such that $|I_1(t, x, y)| \leq \varepsilon/3$, for any $x, y \in H$ such that $|x-y| \leq \delta_1$. For $I_2(t, x, y)$ we have, by elementary inequalities,

$$|I_2(t, x, y)| \leq \frac{\|\varphi\|_{0,1}}{1+|x|}\mathbb{E}[||X(t, x)| - |X(t, y)||]$$

$$\leq \frac{\|\varphi\|_{0,1}}{1+|x|}\mathbb{E}[|X(t, x) - X(t, y)|] \leq \|\varphi\|_{0,1}c|x - y|.$$

Then there exists $\delta_2 > 0$ such that $|I_2(t, x, y)| \leq \varepsilon/3$, for any $x, y \in H$ such that $|x - y| \leq \delta_2$. Similarly, for $I_3(t, x, y)$ we have

$$|I_3(t, x, y)| \leq \|\varphi\|_{0,1}\frac{1 + \mathbb{E}\left[|X(t, x)|\right]}{1 + |x|}\frac{||x| - |y||}{1 + |y|}$$
$$\leq c\|\varphi\|_{0,1}(1 + c)|x - y|,$$

for some $c > 0$. Then, there exists $\delta_3 > 0$ such that $|I_3(t, x, y)| \leq \varepsilon/3$, for any $x, y \in H$ such that $|x - y| \leq \delta_3$. Finally, for any $x, y \in H$ with $|x - y| \leq \min\{\delta_1, \delta_2, \delta_3\}$ we find that

$$\left|\frac{P_t\varphi(x)}{1 + |x|} - \frac{P_t\varphi(y)}{1 + |y|}\right| < \varepsilon$$

as claimed. Now, by taking into account (5.4), there exists $c > 0$ such that

$$\left|\frac{P_t\varphi(x)}{1 + |x|}\right| \leq \|\varphi\|_{0,1}\frac{1 + \mathbb{E}\left[|X(t, x)|\right]}{1 + |x|} \leq c\|\varphi\|_{0,1}.$$

Then $P_t\varphi \in C_{b,1}(H)$. Note that by (5.4) it follows that the operators P_t are bounded in a neighborhood of 0. Hence, the existence of the two constants $c_0 > 0$, $\omega_0 \in \mathbb{R}$ follows by (iv) and by a standard argument. Notice that by the same argument follows[1] (v).

(ii). Take $\varphi \in C_{b,1}(H)$, and consider a sequence $(\varphi_n)_{n\in\mathbb{N}} \subset C_b(H)$ such that

$$\lim_{n\to\infty} \frac{\varphi_n}{1 + |\cdot|} \stackrel{\pi}{=} \frac{\varphi}{1 + |\cdot|}. \tag{5.11}$$

Since $\pi_t(t, \cdot)$ is the image measure of $X(t, x)$ in H, the representation (ii) holds for any φ_n, that is

$$P_t\varphi_n(x) = \mathbb{E}\left[\varphi_n(X(t, x))\right] = \int_H \varphi_n(y)\pi_t(x, dy).$$

Since (5.4) holds we have $\pi(x, \cdot) \in \mathcal{M}_1(H)$, and by (5.11) there exists $c > 0$ such that $|\varphi_n(x)| \leq c(1 + |x|)$, for any $n \in \mathbb{N}, x \in H$. Finally, the result follows by the dominated convergence theorem.

(iii). For any $\varphi \in C_{b,1}(H), x \in H, t, s \geq 0$ we have

$$P_t\varphi(x) - P_s\varphi(x) = \mathbb{E}\left[\frac{\varphi(X(t, x))}{1 + |X(t, x)|} - \frac{\varphi(X(s, x))}{1 + |X(s, x)|}(1 + |X(t, x)|)\right]$$
$$+ \mathbb{E}\left[\frac{\varphi(X(s, x))}{1 + |X(s, x)|}(|X(t, x)| - |X(s, x)|)\right].$$

---

[1] Of course, to prove (iv)-(v) we do not use this statement of (i).

Then

$$|P_t\varphi(x) - P_s\varphi(x)| \le \mathbb{E}\left[\theta_\varphi\left(|X(t,x) - X(s,x)|\right)\left(1 + |X(t,x)|\right)\right]$$
$$+ \|\varphi\|_{0,1}\mathbb{E}\left[|X(t,x) - X(s,x)|\right], \quad (5.12)$$

where $\theta_\varphi : \mathbb{R}^+ \to \mathbb{R}$ is the modulus of continuity of $(1 + |\cdot|)^{-1}\varphi$. Note also that since for any $x \in H$ the process $(X(t,x))_{t\ge 0}$ is continuous in mean square, we have

$$\lim_{t\to s}|X(t,x) - X(s,x)| = 0 \quad \mathbb{P}\text{-a.s..}$$

Hence, by taking into account that $\theta_\varphi : \mathbb{R}^+ \to \mathbb{R}$ is bounded and that (5.4) holds, we can apply the dominated convergence theorem to show that the first term in the right-hand side of (5.12) vanishes as $t \to s$. Finally, the fact that the second term on the right-hand side of (5.12) vanishes as $t \to s$ may be proved by the same argument.

(iv). Take $\varphi \in C_{b,1}(H)$, and consider a sequence $(\varphi_n)_{n\in\mathbb{N}} \subset C_b(H)$ such that $(1 + |\cdot|)^{-1}\varphi_n \xrightarrow{\pi} (1 + |\cdot|)^{-1}\varphi$ as $n \to \infty$. By the markovianity of the process $X(t,x)$ it follows that (iv) holds true for any φ_n. Then, since by (iii) $(1 + |\cdot|)^{-1}P_t\varphi_n \xrightarrow{\pi} (1 + |\cdot|)^{-1}P_t\varphi$ as $n \to \infty$, still by (iii) we find

$$\frac{P_{t+s}\varphi}{1 + |\cdot|} \overset{\pi}{=} \lim_{n\to\infty}\frac{P_{t+s}\varphi_n}{1 + |\cdot|} = \lim_{n\to\infty}\frac{P_t P_s\varphi_n}{1 + |\cdot|} \overset{\pi}{=} \frac{P_t P_s\varphi}{1 + |\cdot|}.$$

This concludes the proof.   □

Remark 5.1.7. *We recall that for any $k > 0$, $T > 0$ there exists $c_k > 0$ such that*

$$\sup_{t\in[0,T]} \mathbb{E}\left[|X(t,x)|^k\right] < c_k\left(1 + |x|^k\right),$$

that implies $\{\pi_t(x,\cdot),\ t \ge 0,\ x \in H\} \subset \bigcap_{k\ge 0}\mathcal{M}_k(H)$. Consequently, all the results of this section are true with $C_{b,k}(H)$ replacing $C_{b,1}(H)$.

Here we collect some properties of the generator $(K, D(K, C_{b,1}(H)))$.

Proposition 5.1.8. *Let $X(t,x)$ be the mild solution of problem (5.1) and let $(P_t)_{t\ge 0}$ be the associated transition semigroups in the space $C_{b,1}(H)$ defined by (5.5). Let also $(K, D(K, C_{b,1}(H)))$ be the associated infinitesimal generators, defined by (5.6). Then*

(i) *for any $\varphi \in D(K, C_{b,1}(H))$, we have $P_t\varphi \in D(K, C_{b,1}(H))$ and $KP_t\varphi = P_t K\varphi,\ t \ge 0$;*

(ii) *for any $\varphi \in D(K, C_{b,1}(H))$, $x \in H$, the map $[0,\infty) \to \mathbb{R}$, $t \mapsto P_t\varphi(x)$ is continuously differentiable and $(d/dt)P_t\varphi(x) = P_t K\varphi(x)$;*

(iii) *given $c_0 > 0$ and ω_0 as in Proposition 5.1.6, for any $\lambda > \omega_0$ the linear operator $R(\lambda, K)$ on $C_{b,1}(H)$ doefined by*

$$R(\lambda, K)f(x) = \int_0^\infty e^{-\lambda t} P_t f(x)dt, \quad f \in C_{b,1}(H), \ x \in H$$

satisfies, for any $f \in C_{b,1}(H)$

$$R(\lambda, K) \in \mathcal{L}(C_{b,1}(H)), \qquad \|R(\lambda, K)\|_{\mathcal{L}(C_{b,1}(H))} \le \frac{c_0}{\lambda - \omega_0}$$

$$R(\lambda, K)f \in D(K, C_{b,1}(H)), \quad (\lambda I - K)R(\lambda, K)f = f.$$

We call $R(\lambda, K)$ the resolvent *of K at λ;*
(iv) *for any $\varphi \in C_{b,1}(H)$, $t > 0$, the function*

$$H \to \mathbb{R}, \quad x \mapsto \int_0^t P_s\varphi(x)ds$$

belongs to $D(K, C_{b,1}(H))$, and it holds

$$K\left(\int_0^t P_s\varphi ds\right) = P_t\varphi - \varphi.$$

Proof. (i). It is proved by taking into account (5.6) and (iii) of Proposition 5.1.6.
(ii). This follows easily by (i) and by (iii) of Proposition 5.1.6.
(iii). By (i) of Proposition 5.1.6 we have

$$\left\|\int_0^\infty e^{-\lambda t} P_t f dt\right\|_{0,1} \le c_0 \int_0^\infty e^{-(\lambda-\omega_0)t} dt \|f\|_{0,1} = \frac{c_0 \|f\|_{0,1}}{\lambda - \omega_0}.$$

Finally, the fact that $R(\lambda, K)f \in D(K, C_{b,1}(H))$ and $(\lambda I - K)R(\lambda, K)f = f$ hold can be proved in a standard way (see, for instance, [8, 40]).
(iv). The proof is the same of Theorem 2.2.4. $\qquad\square$

5.2. Proof of Thcorem 5.1.2

In order to prove this theorem, we need some results about the transition semigroup $(P_t)_{t\ge 0}$ in the space $C_b(H)$. Since for any $\varphi \in C_b(H)$ the representation formula

$$P_t\varphi(x) = \int_H \varphi(y)\pi_t(x, dy), \quad x \in H, \ t \ge 0$$

holds (*cf.* (ii) of Proposition 5.1.6) and $X(t, x)$ is continuous in mean square, it follows easily that $(P_t)_{t\geq0}$ is a semigroup of contraction operators in the space $C_b(H)$. Moreover, we have that for any $x \in H$, $\varphi \in C_b(H)$ the function $\mathbb{R}^+ \to \mathbb{R}$, $t \mapsto P_t\varphi(x)$ is continuous (*cf.* (iii) of Proposition 5.1.6). This means that $(P_t)_{t\geq0}$ fulfills Definition 2.2.1, namely it is a stochastically continous Markov semigroup.

Following (6), we denote by $(K, D(K, C_b(H)))$ the infinitesimal generator of P_t is the space $C_b(H)$, defined by

$$
\begin{cases}
D(K, C_b(H)) = \left\{ \varphi \in C_b(H) : \exists g \in C_b(H), \lim_{t \to 0^+} \dfrac{P_t\varphi(x) - \varphi(x)}{t} = g(x), \right. \\
\left. \quad x \in H, \ \sup_{t \in (0,1)} \left\| \dfrac{P_t\varphi - \varphi}{t} \right\|_0 < \infty \right\} \\[2mm]
K\varphi(x) = \lim_{t \to 0^+} \dfrac{P_t\varphi(x) - \varphi(x)}{t}, \quad \varphi \in D(K, C_b(H)), \ x \in H.
\end{cases}
$$
(5.13)

It is clear that $D(K, C_b(H)) \subset D(K, C_{b,1}(H))$. Hence, by applying Theorem 2.3.1 to $(K, D(K, C_b(H)))$ it yields

Theorem 5.2.1. *For any $\mu \in \mathcal{M}(H)$ there exists a unique family of measures $\{\mu_t, \ t \geq 0\} \subset \mathcal{M}(H)$ such that*

$$
\int_0^T |\mu_t|_{TV}(H)dt < \infty, \quad \forall T > 0
$$
(5.14)

and

$$
\int_H \varphi(x)\mu_t(dx) - \int_H \varphi(x)\mu(dx) = \int_0^t \left(\int_H K\varphi(x)\mu_s(dx) \right) ds
$$
(5.15)

holds for any $t \geq 0$, $\varphi \in D(K, C_b(H))$.

We split the proof into two lemmata.

Lemma 5.2.2. *The formula*

$$
\langle \varphi, P_t^* F \rangle_{\mathcal{L}(C_{b,1}(H),(C_{b,1}(H))^*)} = \langle P_t\varphi, F \rangle_{\mathcal{L}(C_{b,1}(H),(C_{b,1}(H))^*)}
$$
(5.16)

defines a semigroup of linear operators in $(C_{b,1}(H))^*$. Finally, $P_t^* : \mathcal{M}_1(H) \to \mathcal{M}_1(H)$ and it maps positive measures into positive measures.

Proof. Fix $t \geq 0$. By (5.4) it follows that there exists $c > 0$ such that $|P_t \varphi(x)| \leq c \|\varphi\|_{0,1}(1 + |x|)$, for any $\varphi \in C_{b,1}(H)$. Then, if $F \in (C_{b,1}(H))^*$, we have

$$\left| \langle \varphi, P_t^* F \rangle_{\mathcal{L}(C_{b,1}(H),(C_{b,1}(H))^*)} \right| \leq c \|F\|_{(C_{b,1}(H))^*} \|\varphi\|_{0,1},$$

for any $\varphi \in C_{b,1}(H)$. Since P_t^* is linear, it follows that $P_t^* \in \mathcal{L}((C_{b,1}(H))^*)$. Note that by (ii) of Proposition 5.1.6 it follows $P_t \varphi \geq 0$ for any $\varphi \geq 0$. This implies that if $\langle \varphi, F \rangle \geq 0$ for any $\varphi \geq 0$, then $\langle \varphi, P_t^* F \rangle \geq 0$ for any $\varphi \geq 0$. Hence, in order to check that $P_t^* : \mathcal{M}_1(H) \to \mathcal{M}_1(H)$, it is sufficient to take μ positive. So, let $\mu \in \mathcal{M}_1(H)$ be positive and consider the map

$$\Lambda : \mathcal{B}(H) \to \mathbb{R}, \quad \Gamma \mapsto \Lambda(\Gamma) = \int_H \pi_t(x, \Gamma) \mu(dx).$$

We recall that since $X(t, x)$ is continuous with respect to x, for any $\Gamma \in \mathcal{B}(H)$ the map $H \to [0, 1]$, $x \to \pi_t(x, \Gamma)$ is Borel, and consequently the above formula in meaningful. It is straightforward to see that Λ is a positive and finite Borel measure on H, namely $\Lambda \in \mathcal{M}(H)$. We now show $\Lambda = P_t^* \mu$.

Let us fix $\varphi \in C_b(H)$, and consider a sequence of simple Borel functions $(\varphi_n)_{n \in \mathbb{N}}$ which converges uniformly to φ and such that $|\varphi_n(x)| \leq |\varphi(x)|$, $x \in H$. For any $x \in H$ we have

$$\lim_{n \to \infty} \int_H \varphi_n(y) \pi_t(x, dy) = \int_H \varphi(y) \pi_t(x, dy) = P_t \varphi(x)$$

and

$$\sup_{x \in H} \left| \int_H \varphi_n(y) \pi_t(x, dy) \right| \leq \|\varphi\|_0.$$

Hence, by the dominated convergence theorem and by taking into account that φ_n is simple, we have

$$\int_H \varphi(x) \Lambda(dx) = \lim_{n \to \infty} \int_H \varphi_n(x) \Lambda(dx)$$

$$= \lim_{n \to \infty} \int_H \left(\int_H \varphi_n(y) \pi_t(x, dy) \right) \mu(dx)$$

$$= \int_H \left(\int_H \varphi(y) \pi_t(x, dy) \right) \mu(dx) = \int_H P_t \varphi(x) \mu(dx).$$

This implies that $P_t^* \mu = \Lambda$ and consequently $P_t^* \mu \in \mathcal{M}(H)$.

In order to show that $P_t^* \mu \in \mathcal{M}_1(H)$, consider a sequence of functions $(\psi_n)_{n \in \mathbb{N}} \subset C_b(H)$ such that $\psi_n(x) \to |x|$ as $n \to \infty$ and $\psi(x) \leq |x|$, for any $x \in H$. By Proposition 5.1.6 we have $\int_H \psi_n(y)\pi_t(x, dy) \to \int_H |y|\pi_t(x, dy)$ as $n \to \infty$ and $\int_H \psi_n(y)\pi_t(x, dy) \le c(1 + |x|)$, for any $x \in H$ and for some $c > 0$. Hence, since $\mu \in \mathcal{M}_1(H)$ we have

$$\int_H |x| P_t^* \mu(dx) = \lim_{n \to \infty} \int_H \psi_n(x) P_t^* \mu(dx)$$

$$= \lim_{n \to \infty} \int_H \left(\int_H \psi_n(y)\pi_t(x, dy) \right) \mu(dx) \le \int_H c(1 + |x|)\mu(dx) < \infty.$$

This concludes the proof. $\qquad\qquad\qquad\qquad\qquad\qquad\qquad\qquad\square$

Lemma 5.2.3. *For any $\mu \in \mathcal{M}_1(H)$ there exists a unique family of finite Borel measures $\{\mu_t, \ t \ge 0\} \subset \mathcal{M}_1(H)$ fulfilling (5.7), (5.8), and this family is given by $P_t^* \mu, \ t \ge 0$.*

Proof. We first check that $P_t^* \mu, \ t \ge 0$ satisfies (5.7), (5.8). By Proposition 5.2.2, for any $\mu \in \mathcal{M}_1(H)$, formula (5.16) defines a family $\{P_t^* \mu, \ t \ge 0\}$ of measures on H. Moreover, by (i) of Proposition 5.1.6 it follows that for any $T > 0$ it holds

$$\sup_{t \in [0,T]} \| P_t^* \mu \|_{(C_{b,1}(H))^*} = \sup_{t \in [0,T]} \int_H (1 + |x|)|P_t^* \mu|_{TV}(dx) < \infty.$$

Hence, (5.7) holds. We now show (5.8). By (i), (ii), (iv) of Proposition 5.1.6 and by the dominated convergence theorem it follows easily that for any $\varphi \in C_{b,1}(H)$ the function

$$\mathbb{R}^+ \to \mathbb{R}, \quad t \mapsto \int_H \varphi(x) P_t^* \mu(dx) \qquad\qquad (5.17)$$

is continuous. Clearly, $P_0^* \mu = \mu$. Now we show that if $\varphi \in D(K, C_{b,1}(H))$ then the function (5.17) is differentiable. Indeed, by taking into account (5.6) and (i) of Proposition 5.1.8, for any $\varphi \in D(K, C_{b,1}(H))$ we can apply the dominated convergence theorem to obtain

$$\frac{d}{dt} \int_H \varphi(x) P_t^* \mu(dx)$$

$$= \lim_{h \to 0} \frac{1}{h} \left(\int_H P_{t+h}\varphi(x)\mu(dx) - \int_H P_t\varphi(x)\mu_t(dx) \right)$$

$$= \lim_{h \to 0} \int_H \left(\frac{P_{t+h}\varphi(x) - P_t\varphi(x)}{h} \right) \mu(dx)$$

$$= \lim_{h \to 0} \int_H P_t \left(\frac{P_h\varphi - \varphi}{h} \right)(x)\mu(dx)$$

$$= \int_H \lim_{h \to 0} \left(\frac{P_h\varphi - \varphi}{h} \right)(x) P_t^* \mu(dx) = \int_H K\varphi(x) P_t^* \mu(dx).$$

Then, by arguing as above, it follows that the differential of the mapping defined by (5.17) is continuous. This clearly implies that $P_t^* \mu$, $t \geq 0$ satisfies (5.8). In order to show uniqueness of such a solution, by the linearity of the problem it is sufficient to show that if $\mu = 0$ and $\{\mu_t,\ t \geq 0\} \subset \mathcal{M}_1(H)$ is a solution of equation (5.8), then $\mu_t = 0$, for any $t \geq 0$. Note that equation (5.8) holds in particular for $\varphi \in D(K, D(K, C_{b,1}(H)))$ (cf. (5.13)) and consequently (5.15) holds, for any $\varphi \in D(K, D(K, C_{b,1}(H)))$. Note also that by (5.7) follows that μ_t, $t \geq 0$ fulfils (5.14). Hence, by Theorem 5.2.1, it follows that $\mu_t = 0$, $\forall t \geq 0$. This concludes the proof. $\qquad\square$

5.3. Proof of Theorem 5.1.3

We split the proof in several steps. We start by studying the Ornstein-Uhlenbeck operator in $C_{b,1}(H)$ that is, roughly speaking, the case $F = 0$ in (5.9). In Proposition 5.3.2 we shall prove Theorem 5.1.4 in the case $F = 0$. Then, Corollary 5.3.3 will show that $(K, D(K_0))$ is an extension of K_0 and $K\varphi = K_0\varphi$ for any $\varphi \in \mathcal{E}_A(H)$. In order to complete the proof of the theorem, we shall present several approximation results. Finally, Lemma 5.3.5 will complete the proof.

5.3.1. The Ornstein-Uhlenbeck semigroup in $C_{b,1}(H)$

An important role in what follows will be played by the *Ornstein-Uhlenbeck semigroup* $(R_t)_{t \geq 0}$ in the space $C_{b,1}(H)$, defined by the formula

$$R_t\varphi(x) = \begin{cases} \varphi(x), & t = 0, \\ \displaystyle\int_H \varphi(e^{tA}x + y)N_{Q_t}(dy), & t > 0 \end{cases}$$

where $\varphi \in C_{b,1}(H)$, $x \in H$ and N_{Q_t} is the Gaussian measure of zero mean and covariance operator Q_t (cf. Hypothesis 1.2.1 and Chapter 3). We recall that formula (4.9)

$$R_t\varphi(x) = \mathbb{E}\left[\varphi\left(e^{tA}x + \int_0^t e^{(t-s)A} Bd W(s)\right)\right] \qquad (5.18)$$

holds, for any $t \geq 0$, $\varphi \in C_{b,1}(H)$, $x \in H$. Hence, the Ornstein-Uhlenbeck semigroup $(R_t)_{t \geq 0}$ coincides with the transition semigroup (5.5) in the case $F = 0$ in (5.1). Consequently, $(R_t)_{t \geq 0}$ satisfies stamentes (i)–(v) of Proposition 5.1.6. We recall that (3.3) holds, and consequently $R_t : \mathcal{E}_A(H) \to \mathcal{E}_A(H)$, for any $t \geq 0$. We define the infinitesimal generator $L : D(L, C_{b,1}(H)) \to C_{b,1}(H)$ of $(R_t)_{t \geq 0}$ in $C_{b,1}(H)$ as in (5.6), with L replacing K and R_t replacing P_t.

Theorem 5.3.1. *Let $(P_t)_{t\geq 0}$ be the semigroup (5.5) and let $(R_t)_{t\geq 0}$ be the Ornstein-Uhlenbeck semigroup (5.18). We denote by $(K, D(K, C_{b,1}(H)))$ and by $(L, D(L, C_{b,1}(H)))$ the corresponding infinitesimal generators in $C_{b,1}(H)$. Then we have $D(L, C_{b,1}(H)) \cap C_b^1(H) = D(K, C_{b,1}(H)) \cap C_b^1(H)$ and $K\varphi = L\varphi + \langle D\varphi, F\rangle$, for any $\varphi \in D(L, C_{b,1}(H)) \cap C_b^1(H)$.*

Proof. Let $X(t, x)$ be the mild solution of equation (5.1) and set

$$Z_A(t, x) = e^{tA}x + \int_0^t e^{(t-s)A} B dW(s).$$

Take $\varphi \in D(L, C_{b,1}(H)) \cap C_b^1(H)$. By taking into account that

$$X(t, x) = Z_A(t, x) + \int_0^t e^{(t-s)A} F(X(s, x))ds,$$

by the Taylor formula we have that $\mathbb{P}$-a.s. it holds

$$\varphi(Z_A(t, x)) = \varphi(Z_A(t, x)) - \varphi(X(t, x)) + \varphi(X(t, x)) = \varphi(X(t, x))$$

$$- \int_0^1 \left\langle D\varphi(\xi Z_A(t, x) + (1 - \xi)X(t, x)), \int_0^t e^{(t-s)A} F(X(s, x))ds \right\rangle d\xi.$$

Then we have

$$R_t\varphi(x) - \varphi(x) = \mathbb{E}\big[\varphi(Z_A(t, x))\big] - \varphi(x) = P_t\varphi(x) - \varphi(x)$$

$$- \mathbb{E}\left[\int_0^1 \left\langle D\varphi(\xi Z_A(t, x) + (1-\xi)X(t, x)), \int_0^t e^{(t-s)A} F(X(s, x))ds \right\rangle d\xi \right].$$

Before proceeding, we need the following

Claim. For any $x \in H$ it holds

$$\lim_{t\to 0^+} \frac{1}{t} \mathbb{E}\left[\int_0^1 \left\langle D\varphi(\xi Z_A(t, x) + (1-\xi)X(t, x)), \int_0^t e^{(t-s)A} F(X(s, x))ds \right\rangle d\xi \right]$$

$$= \langle D\varphi(x), F(x)\rangle. \tag{5.19}$$

For any $x \in H$ we have

$$
\frac{1}{t}\mathbb{E}\left[\int_0^1 \left\langle D\varphi(\xi Z_A(t,x)+(1-\xi)X(t,x)), \int_0^t e^{(t-s)A}F(X(s,x))ds\right\rangle d\xi\right]
$$
$$
- \langle D\varphi(x), F(x)\rangle
$$
$$
= \mathbb{E}\left[\int_0^1 \langle D\varphi(\xi Z_A(t,x)+(1-\xi)X(t,x)) - D\varphi(x), F(x)\rangle d\xi\right]
$$
$$
+ \frac{1}{t}\mathbb{E}\left[\int_0^1 \left\langle D\varphi(\xi Z_A(t,x)+(1-\xi)X(t,x)),\right.\right.
$$
$$
\left.\left. \int_0^t e^{(t-s)A}F(X(s,x))ds - F(x)\right\rangle d\xi\right]
$$
$$
= I_1(t,x) + I_2(t,x).
$$

For $I_1(t,x)$ we have

$$
|I_1(t,x)| \leq \mathbb{E}\left[\int_0^1 |D\varphi(\xi Z_A(t,x)+(1-\xi)X(t,x)) - D\varphi(x)| d\xi\right]|F(x)|
$$
$$
\leq c_F \mathbb{E}\left[\int_0^1 \theta_{D\varphi}(|\xi Z_A(t,x)+(1-\xi)X(t,x)-x|)d\xi\right](1+|x|)
$$

where $\theta_{D\varphi}: \mathbb{R}^+ \to \mathbb{R}^+$ is the modulus of continuity of $D\varphi$ and $c_F > 0$ is such that $|F(x)| \leq c_F(1+|x|), \forall x \in H$. Since $\mathbb{E}[|Z_A(t,x) - x|^2] \to 0$ and $\mathbb{E}[|X(t,x) - x|^2] \to 0$ as $t \to 0^+$, it follows

$$
\lim_{t\to 0^+} I_1(t,x) = 0, \quad \forall x \in H.
$$

For $I_2(t,x)$ we have

$$
|I_2(t,x)| \leq \|D\varphi\|_{C_b(H;H)}\mathbb{E}\left[\frac{1}{t}\left|\int_0^t e^{(t-s)A}(F(X(s,x)) - F(x))ds\right|\right]
$$
$$
+ \|D\varphi\|_{C_b(H;H)}\left|\frac{1}{t}\int_0^t e^{(t-s)A}F(x)ds - F(x)\right|
$$
$$
= I_{2,1}(t,x) + I_{2,2}(t,x).
$$

Notice that by Hypothesis 1.2.1

$$
I_{2,1}(t,x) \leq \frac{M}{t}\int_0^t e^{(t-s)\omega}\mathbb{E}[|F(X(s,x)) - F(x)|]ds
$$
$$
\leq \kappa\frac{M}{t}\int_0^t e^{(t-s)\omega}\mathbb{E}[|X(s,x) - x|]ds.
$$

Consequently, since $\mathbb{E}[|X(t, x) - x|^2] \to 0$ as $t \to 0^+$, it follows that

$$\lim_{t \to 0^+} I_{2,1}(t, x) \to 0.$$

For $I_{2,2}(t, x)$ we have, by the fact that the semigroup e^{tA}, $t \geq 0$ is strongly continuous,

$$\lim_{t \to 0^+} I_{2,2}(t, x) \to 0.$$

Then,

$$\lim_{t \to 0^+} I_2(t, x) = 0, \quad \forall x \in H.$$

This prove the claim.

By taking into account that $\varphi \in D(L, C_{b,1}(H)) \cap C_b^1(H)$ and that (5.19) holds, for any $x \in H$ we have

$$\lim_{t \to 0^+} \frac{P_t \varphi(x) - \varphi(x)}{t} = L\varphi(x) + \langle D\varphi(x), F(x) \rangle.$$

As easily seen, $x \mapsto L\varphi(x) + \langle D\varphi(x), F(x) \rangle$ is uniformly continuous. Moreover, since $t \to \mathbb{E}[|X(t, x)|]$ is continuous and $\mathbb{E}[|X(t, x) - x|] \to 0$ as $t \to 0^+$, there exists $c > 0$ such that

$$\left| \frac{P_t \varphi(x) - \varphi(x)}{t} \right| \leq c(1 + |x|)$$

$$+ c_F \|D\varphi\|_{C_b(H;H)} \frac{M}{t} \int_0^t e^{(t-s)\omega}(1 + \mathbb{E}[|X(s, x)|])ds$$

$$\leq c \left(1 + c_F \|D\varphi\|_{C_b(H;H)} \frac{M}{t} \int_0^t e^{(t-s)\omega}ds \right) (1 + |x|).$$

This implies

$$\sup_{t \in (0,1]} \left\| \frac{P_t \varphi - \varphi}{t} \right\|_{0,1} < \infty.$$

Hence, $\varphi \in D(K, C_{b,1}(H)) \cap C_b^1(H)$ and $K\varphi = L\varphi + \langle D\varphi, F \rangle$ as claimed. The opposite inclusion follows by interchanging the role of R_t and P_t in the Taylor formula. $\quad\square$

Let us set

$$L_0\varphi(x) = \frac{1}{2}\text{Tr}[BB^* D^2\varphi(x)] + \langle x, A^* D\varphi(x) \rangle, \quad \varphi \in \mathcal{E}_A(H), \ x \in H.$$

We need the following approximation result:

Proposition 5.3.2. *For any $\varphi \in \mathcal{E}_A(H)$ we have $\varphi \in D(L, C_{b,1}(H))$ and*

$$L\varphi = L_0\varphi. \tag{5.20}$$

The set $\mathcal{E}_A(H)$ is a π-core for the infinitesimal generator $(L, D(L, C_{b,1}(H)))$, and for any $\varphi \in D(L, C_{b,1}(H)) \cap C_b^1(H)$ there exists $m \in \mathbb{N}$ and an m-indexed sequence $(\varphi_{n_1,\ldots,n_m})_{n_1,\ldots,n_m \in \mathbb{N}} \subset \mathcal{E}_A(H)$ such that

$$\lim_{n_1 \to \infty} \cdots \lim_{n_m \to \infty} \frac{\varphi_{n_1,\ldots,n_m}}{1 + |\cdot|} \stackrel{\pi}{=} \frac{\varphi}{1 + |\cdot|}, \tag{5.21}$$

$$\lim_{n_1 \to \infty} \cdots \lim_{n_m \to \infty} \frac{L_0\varphi_{n_1,\ldots,n_m}}{1 + |\cdot|} \stackrel{\pi}{=} \frac{L\varphi}{1 + |\cdot|}. \tag{5.22}$$

Finally, if $\varphi \in D(L, C_{b,1}(H)) \cap C_b^1(H)$ we can choose the sequence in such a way that (5.21), (5.22) hold and

$$\lim_{n_1 \to \infty} \cdots \lim_{n_m \to \infty} \langle D\varphi_{n_1,\ldots,n_m}, h \rangle \stackrel{\pi}{=} \langle D\varphi, h \rangle, \tag{5.23}$$

for any $h \in H$.

Proof. We recall that the proof of (5.20) may be found in [13], Remark 2.66 (in [13] the result is proved for the semigroup $(R_t)_{t \geq 0}$ in the space $C_{b,2}(H)$, but it is clear that the result holds also in $C_{b,1}(H)$).

Here we give only a sketch of the proof, which is very similar to the proof given in [37]. Take $\varphi \in D(L, C_{b,1}(H))$. For any $n_2 \in \mathbb{N}$, set

$$\varphi_{n_2}(x) = \frac{n_2\varphi(x)}{n_2 + |x|^2}.$$

Clearly, $\varphi_{n_2} \in C_b(H)$ and $(1 + |\cdot|)^{-1}\varphi_{n_2} \stackrel{\pi}{\to} (1 + |\cdot|)^{-1}\varphi$ as $n_2 \to \infty$. By Proposition 3.1.3, for any $n_2 \in \mathbb{N}$ we fix a sequence[2] $(\varphi_{n_2,n_3})_{n_3 \in \mathbb{N}} \subset \mathcal{E}_A(H)$ such that $\varphi_{n_2,n_3} \stackrel{\pi}{\to} \varphi_{n_2}$ as $n_3 \to \infty$. Set now, for any $n_1, n_2, n_3, n_4 \in \mathbb{N}$

$$\varphi_{n_1,n_2,n_3,n_4}(x) = \frac{1}{n_4} \sum_{k=1}^{n_4} R_{\frac{k}{n_1 n_4}} \varphi_{n_2 n_3}(x). \tag{5.24}$$

Since for any $\varphi \in C_{b,1}(H)$, $x \in H$ the function $\mathbb{R}^+ \to \mathbb{R}$, $t \mapsto R_t\varphi(x)$ is continuous, a straightforward computation shows that the sequence

---

[2] We assume that the sequence has only one index.

$(\varphi_{n_1,...,n_4})$ fulfils (5.21). Similarly, we find that for any $x \in H$ it holds

$$\lim_{n_1\to\infty} \lim_{n_2\to\infty} \lim_{n_3\to\infty} \lim_{n_4\to\infty} L_0\varphi_{n_1,n_2,n_3,n_4}(x)$$

$$= \lim_{n_1\to\infty} \lim_{n_2\to\infty} \lim_{n_3\to\infty} \lim_{n_4\to\infty} L\varphi_{n_1,n_2,n_3,n_4}(x)$$

$$= \lim_{n_1\to\infty} \lim_{n_2\to\infty} \lim_{n_3\to\infty} n_1 \int_0^{\frac{1}{n_1}} LR_t\varphi_{n_2,n_3}(x)dt$$

$$= \lim_{n_1\to\infty} \lim_{n_2\to\infty} \lim_{n_3\to\infty} n_1 \left(R_{\frac{1}{n_1}}\varphi_{n_2,n_3}(x) - \varphi_{n_2,n_3}(x) \right)$$

$$= \lim_{n_1\to\infty} \left(R_{\frac{1}{n_1}}\varphi(x) - \varphi(x) \right) = L\varphi(x).$$

Here we have used the continuity of $t \mapsto LR_t\varphi_{n_2,n_3}(x)$ and the fact that $LR_t\varphi_{n_2,n_3}(x) = (d/dt)R_t\varphi_{n_2,n_3}(x)$ (cf. Proposition 5.1.6 and Proposition 5.1.8). Let us check that any of the above limit is equibounded in $C_{b,1}(H)$ with respect to the corresponding index. By (5.24) we have

$$\sup_{n_4\in\mathbb{N}} \|L\varphi_{n_1,n_2,n_3,n_4}\|_{0,1} \le \|L\varphi_{n_2n_3}\|_0.$$

By contruction of $(\varphi_{n_2,n_3})_{n_2,n_3}$ we have

$$\sup_{n_3\in\mathbb{N}} \left\| n_1 \left(R_{\frac{1}{n_1}}\varphi_{n_2,n_3} - \varphi_{n_2,n_3} \right) \right\|_{0,1} \le \sup_{n_3\in\mathbb{N}} \|2n_1\varphi_{n_2,n_3}\|_0 < \infty$$

and

$$\sup_{n_2\in\mathbb{N}} \left\| n_1 \left(R_{\frac{1}{n_1}}\varphi_{n_2} - \varphi_{n_2} \right) \right\|_{0,1} \le \|2n_1\varphi\|_{0,1}.$$

Finally,

$$\sup_{n_1\in\mathbb{N}} \left\| \left(R_{\frac{1}{n_1}}\varphi - \varphi \right) \right\|_{0,1} < \infty$$

since $\varphi \in D(L, C_{b,1}(H))$. Hence, (5.22) follows.

If $\varphi \in D(L, C_{b,1}(H)) \cap C_b^1(H)$, by Proposition 3.1.3, there exists a sequence[3] $(\varphi_n)_{n\in\mathbb{N}} \subset \mathcal{E}_A(H)$ such that $\varphi_n \xrightarrow{\pi} \varphi$ as $n \to \infty$ and $\langle D\varphi_n, h \rangle \xrightarrow{\pi} \langle D\varphi, h \rangle$ as $n \to \infty$, for any $h \in H$. Since for any $t > 0$, $n \in \mathbb{N}$ we have

$$\langle DR_t\varphi_n(x), h \rangle = \int_H \langle D\varphi_n(e^{tA}x + y), h \rangle N_{Q_t}(dy), \quad x \in H$$

it follows $\langle DR_t\varphi_n, h \rangle \xrightarrow{\pi} \langle DR_t\varphi, h \rangle$ as $n \to \infty$, for any $h \in H$. Then, the claim follows by arguing as above. $\square$

---

[3] We assume that the sequence has only one index.

By Theorem 5.3.1 and Proposition 5.3.2 we have:

Corollary 5.3.3. $(K, D(K, C_{b,1}(H)))$ *is an extension of K_0 and for any $\varphi \in \mathcal{E}_A(H)$ we have $\varphi \in D(K, C_{b,1}(H))$ and $K\varphi = K_0\varphi$.*

5.3.2. Approximation of F with smooth functions

It is convenient to introduce an auxiliary Ornstein–Uhlenbeck semigroup

$$U_t\varphi(x) = \int_H \varphi(e^{tS}x + y)N_{\frac{1}{2}S^{-1}(e^{2tS}-1)}(dy), \quad \varphi \in C_b(H)$$

where $S : D(S) \subset H \to H$ is a self–adjoint negative definite operator such that S^{-1} is of trace class. We notice that U_t is strong Feller, and for any $t > 0$, $\varphi \in C_b(H)$, $U_t\varphi$ is infinite times differentiable with bounded differentials (see [13]). We introduce a regularization of F by setting

$$\langle F_n(x), h \rangle = \int_H \left\langle F\left(e^{\frac{1}{n}S}x + y\right), e^{\frac{1}{n}S}h \right\rangle N_{\frac{1}{2}S^{-1}(e^{\frac{2}{n}S}-1)}(dy), \quad n \in \mathbb{N}.$$

It is easy to check that F_n is infinite times differentiable, with first differential bounded by κ, for any $n \in \mathbb{N}$. Moreover, $F_n(x) \to F(x)$ as $n \to \infty$ for all $x \in H$ and $|F_n(x)| \le |F(x)|$, for all $n \in \mathbb{N}$, $x \in H$.

Let P_t^n be the transition semigroup

$$P_t^n\varphi(x) = \mathbb{E}[\varphi(X^n(t, x))], \quad \varphi \in C_{b,1}(H) \tag{5.25}$$

where $X^n(t, x)$ is the solution of (5.1) with F_n replacing F. It is easy to check

$$\lim_{n\to\infty} \mathbb{E}\left[|X^n(t, x) - X(t, x)|^2\right] = 0, \quad t \ge 0, \; x \in H$$

and

$$\mathbb{E}\left[|X^n(t, x)|\right] \le \mathbb{E}\left[|X(t, x)|\right], \quad t \ge 0, \; x \in H,$$

where $c_0 > 0$, $\omega_0 \in \mathbb{R}$ are as in Proposition 5.1.6. This implies

$$\lim_{n\to\infty} \frac{P_t^n\varphi}{1+|\cdot|} \overset{\pi}{=} \frac{P_t\varphi}{1+|\cdot|}, \tag{5.26}$$

for any $t \ge 0$, $\varphi \in C_{b,1}(H)$. We denote by $(K_n, D(K_n, C_{b,1}(H)))$ the infinitesimal generator of the semigroup P_t^n in $C_{b,1}(H)$, defined as in (5.6) with K_n replacing K and P_t^n replacing P_t. We recall that all the statements of Proposition 5.1.6, Theorem 5.2.1 hold also for P_t^n and $(K_n, D(K_n, C_{b,1}(H)))$. We also recall that the resolvent of $(K, D(K, C_{b,1}(H)))$ in $C_{b,1}(H)$ is defined for any $\lambda > \omega_0$ by the formula

$R(\lambda, K) f(x) = \int_0^\infty e^{-\lambda t} P_t f(x) dt$, $f \in C_{b,1}(H)$, $x \in H$ (cf. Theorem 5.2.1). Similarly, for a fixed $n \in \mathbb{N}$ the resolvent of $(K_n, D(K_n, C_{b,1}(H)))$ in $C_{b,1}(H)$ at $\lambda > 0$ is defined by the same formula with P_t^n replacing P_t. Since (5.26) holds, it is straightforward to see that

$$\lim_{n \to \infty} \frac{R(\lambda, K_n)\varphi}{1 + |\cdot|} \stackrel{\pi}{=} \frac{R(\lambda, K)\varphi}{1 + |\cdot|}, \qquad (5.27)$$

for any $\varphi \in C_{b,1}(H)$, $\lambda > \omega_0$.

The following proposition follows by Corollary 4.9 of [37] and by the fact that $\|DF_n\| \leq \kappa$, for any $n \in \mathbb{N}$.

Proposition 5.3.4. *For any $n \in \mathbb{N}$, let $(K_n, D(K_n, C_{b,1}(H)))$ be the infinitesimal generator of the semigroup (5.25). Then for any $\lambda > \max\{0, \omega + M\kappa\}$, the resolvent $R(\lambda, K_n)$ of K_n at λ maps C_b^1 into $C_b^1(H)$ and it holds*

$$\|DR(\lambda, K_n) f\|_{C_b(H;H)} \leq \frac{M\|Df\|_{C_b(H;H)}}{\lambda - (\omega + M\kappa)}, \qquad f \in C_b^1(H). \qquad (5.28)$$

Corollary 5.3.3 shows that K is an extension of K_0 and that $K\varphi = K_0\varphi$, $\forall \varphi \in \mathcal{E}_A(H)$. So, in view of the fact that $KP_t\varphi = P_t K_0\varphi$ for any $\varphi \in \mathcal{E}_A(H)$ (cf. (i) of Proposition 5.1.8), it is not difficult to check that $P_t^*\mu$, $t \geq 0$ fulfils (5.7), (5.10). Now, let $\mu \in \mathcal{M}_1(H)$ and assume that $\{\mu_t, \ t \geq 0\} \subset \mathcal{M}_1(H)$ fulfils (5.7), (5.10). In view of Theorem 5.2.3, to prove that $\mu_t = P_t^*\mu$, for any $t \geq 0$, it is sufficient to show that $\mu_t, t \geq 0$ is also a solution of (5.8). In order to do this, we need an approximation result.

Lemma 5.3.5. *The set $\mathcal{E}_A(H)$ is a π-core for $(K, D(K, C_{b,1}(H)))$, and for any $\varphi \in D(K, C_{b,1}(H))$ there exist $m \in \mathbb{N}$ and an m-indexed sequence $(\varphi_{n_1,\dots,n_m}) \subset \mathcal{E}_A(H)$ such that*

$$\lim_{n_1 \to \infty} \cdots \lim_{n_m \to \infty} \frac{\varphi_{n_1,\dots,n_m}}{1 + |\cdot|} \stackrel{\pi}{=} \frac{\varphi}{1 + |\cdot|}, \qquad (5.29)$$

$$\lim_{n_1 \to \infty} \cdots \lim_{n_m \to \infty} \frac{K_0\varphi_{n_1,\dots,n_m}}{1 + |\cdot|} \stackrel{\pi}{=} \frac{K\varphi}{1 + |\cdot|}. \qquad (5.30)$$

Proof.

Step 1. Let[4] $\varphi \in D(K, C_{b,1}(H))$, $\lambda > \max\{0, \omega_0, \omega + M\kappa\}$ and set $f = \lambda\varphi - K\varphi$. We fix a sequence $(f_{n_1}) \subset C_b^1(H)$ such that

$$\lim_{n_1 \to \infty} \frac{f_{n_1}}{1 + |\cdot|} \stackrel{\pi}{=} \frac{f}{1 + |\cdot|}.$$

---

[4] The assumpion $\lambda > \omega_0$ is necessary to define the resolvent of K (cf. Proposition 5.1.8).

Set $\varphi_{n_1} = R(\lambda, K) f_{n_1}$. By Proposition 5.1.8 it follows that

$$\lim_{n_1 \to \infty} \frac{\varphi_{n_1}}{1 + |\cdot|} \stackrel{\pi}{=} \frac{\varphi}{1 + |\cdot|}, \qquad \lim_{n_1 \to \infty} \frac{K\varphi_{n_1}}{1 + |\cdot|} \stackrel{\pi}{=} \frac{K\varphi}{1 + |\cdot|}. \tag{5.31}$$

Step 2. Now set $\varphi_{n_1,n_2} = R(\lambda, K_{n_2}) f_{n_1}$, where K_{n_2} is the infinitesimal generator of the semigroup $P_t^{n_2}$, introduced in (5.25). Since $f_{n_1} \in C_b^1(H)$, by Proposition 5.3.4 we have $\varphi_{n_1,n_2} \in C_b^1(H)$ and

$$\sup_{n_2 \in \mathbb{N}} \|D\varphi_{n_1,n_2}\|_{C_b(H;H)} \leq \frac{M \|Df_{n_1}\|_{C_b(H;H)}}{\lambda - (\omega + M\kappa)}, \tag{5.32}$$

for any $n_1 \in \mathbb{N}$. Moreover, by (5.27) it holds

$$\lim_{n_2 \to \infty} \varphi_{n_1,n_2} \stackrel{\pi}{=} \varphi_{n_1}, \qquad \lim_{n_2 \to \infty} K_{n_2}\varphi_{n_1,n_2} \stackrel{\pi}{=} K\varphi_{n_1}, \tag{5.33}$$

for any $n_1 \in \mathbb{N}$. Since $\varphi_{n_1,n_2} \in D(K_{n_2}, C_{b,1}(H)) \cap C_b^1(H)$, by Theorem 5.3.1 we have

$$K_{n_2}\varphi_{n_1,n_2} = L\varphi_{n_1,n_2} + \langle D\varphi_{n_1,n_2}, F_{n_2} \rangle.$$

Step 3. By Proposition 5.3.2, for any $n_1, n_2 \in \mathbb{N}$ there exists a sequence $(\varphi_{n_1,n_2,n_3}) \subset \mathcal{E}_A(H)$ (we still assume that it has only one index) such that

$$\lim_{n_3 \to \infty} \varphi_{n_1,n_2,n_3} \stackrel{\pi}{=} \varphi_{n_1,n_2}, \qquad \lim_{n_3 \to \infty} \frac{L\varphi_{n_1,n_2,n_3}}{1 + |\cdot|} \stackrel{\pi}{=} \frac{L\varphi_{n_1,n_2}}{1 + |\cdot|} \tag{5.34}$$

and

$$\lim_{n_3 \to \infty} \langle D\varphi_{n_1,n_2,n_3}, h \rangle \stackrel{\pi}{=} \langle D\varphi_{n_1,n_2}, h \rangle,$$

for any $h \in H$. Notice that, since F_{n_2} is globally Lipschitz, it follows that

$$\lim_{n_3 \to \infty} \frac{\langle D\varphi_{n_1,n_2,n_3}, F_{n_2} \rangle}{1 + |\cdot|} \stackrel{\pi}{=} \frac{\langle D\varphi_{n_1,n_2}, F_{n_2} \rangle}{1 + |\cdot|}.$$

This, together with (5.34), implies that the sequence (φ_{n_1,n_2,n_3}) fulfils

$$\lim_{n_3 \to \infty} \varphi_{n_1,n_2,n_3} \stackrel{\pi}{=} \varphi_{n_1,n_2}, \qquad \lim_{n_3 \to \infty} \frac{K_{n_2}\varphi_{n_1,n_2,n_3}}{1 + |\cdot|} \stackrel{\pi}{=} \frac{K_{n_2}\varphi_{n_1,n_2}}{1 + |\cdot|}.$$

Since K is an extension of K_0 (cf. Corollary 5.3.3), we have

$$K\varphi_{n_1,n_2,n_3} = K_0\varphi_{n_1,n_2,n_3} = K_{n_2}\varphi_{n_1,n_2,n_3} + \langle D\varphi_{n_1,n_2,n_3}, F - F_{n_2} \rangle$$

for any $n_1, n_2, n_3 \in \mathbb{N}$. So we find

$$\lim_{n_3 \to \infty} \frac{K_0 \varphi_{n_1, n_2, n_3}}{1 + |\cdot|} \stackrel{\pi}{=} \frac{K_{n_2} \varphi_{n_1, n_2} + \langle D\varphi_{n_1, n_2}, F - F_{n_2} \rangle}{1 + |\cdot|}, \qquad (5.35)$$

for any $n_1, n_2 \in \mathbb{N}$. Moreover, by (5.32), we see that

$$\frac{|\langle D\varphi_{n_1, n_2}(x), F(x) - F_{n_2}(x) \rangle|}{1 + |x|} \leq \frac{M \|Df_{n_1}\|_{C_b(H;H)}}{\lambda - (\omega + M\kappa)} \frac{|F(x) - F_{n_2}(x)|}{1 + |x|}$$

and consequently

$$\lim_{n_2 \to \infty} \frac{\langle D\varphi_{n_1, n_2}, F - F_{n_2} \rangle}{1 + |\cdot|} \stackrel{\pi}{=} 0.$$

This, together with (5.33) implies

$$\lim_{n_2 \to \infty} \frac{K_{n_2} \varphi_{n_1, n_2} + \langle D\varphi_{n_1, n_2}, F - F_{n_2} \rangle}{1 + |\cdot|} \stackrel{\pi}{=} \frac{K \varphi_{n_1}}{1 + |\cdot|}. \qquad (5.36)$$

Finally, by taking into account (5.31), (5.35), (5.36), we can conclude that the sequence $(\varphi_{n_1, n_2, n_3})_{n_1, n_2, n_3}$ fulfils

$$\lim_{n_1 \to \infty} \lim_{n_2 \to \infty} \lim_{n_3 \to \infty} \frac{\varphi_{n_1, n_2, n_3}}{1 + |\cdot|} \stackrel{\pi}{=} \frac{\varphi}{1 + |\cdot|},$$

$$\lim_{n_1 \to \infty} \lim_{n_2 \to \infty} \lim_{n_3 \to \infty} \frac{K_0 \varphi_{n_1, n_2, n_3}}{1 + |\cdot|} \stackrel{\pi}{=} \frac{K \varphi}{1 + |\cdot|}.$$

This concludes the proof.   □

5.4. Proof of Theorem 5.1.4

Let $\varphi \in D(K, C_{b,1}(H))$ and assume that $(\varphi_n)_{n \in \mathbb{N}} \subset \mathcal{E}_A(H)$ fulfils (5.29), (5.30) (for simplicity we assume that this sequence has only one index: this does not change the generality of the proof). For any $t \geq 0$ we find

$$\int_H \varphi(x) \mu_t(dx) - \int_H \varphi(x) \mu(dx)$$

$$= \lim_{n \to \infty} \left(\int_H \varphi_n(x) \mu_t(dx) - \int_H \varphi_n(x) \mu(dx) \right)$$

$$= \lim_{n \to \infty} \int_0^t \left(\int_H K_0 \varphi_n(x) \mu_s(dx) \right) ds,$$

since $\varphi_n \in D(K, C_{b,1}(H))$ and $K \varphi_n = K_0 \varphi_n$, for any $n \in \mathbb{N}$ (cf. Corollary 5.3.3). Now observe that since $\sup_{n \in \mathbb{N}} |K_0 \varphi_n(x)| \leq c(1 + |x|)$ for some $c > 0$ and since $\mu_s \in \mathcal{M}_1(H)$ for any $s \geq 0$, it holds

$$\lim_{n \to \infty} \int_H K_0 \varphi_n(x) \mu_s(dx) = \int_H K \varphi(x) \mu_s(dx)$$

and

$$\sup_{n\in\mathbb{N}} \left| \int_H K_0\varphi_n(x)\mu_s(dx) \right| \leq c \int_H (1+|x|)|\mu_s|(dx).$$

Hence, by taking into account (5.7) we can apply the dominated convergence theorem to obtain

$$\lim_{n\to\infty} \int_0^t \left(\int_H K_0\varphi_n(x)\mu_s(dx) \right) ds = \int_0^t \left(\int_H K\varphi(x)\mu_s(dx) \right) ds$$

So, μ_t, $t \geq 0$ is also a solution of the measure equation for $(K, D(K, C_{b,1}(H)))$. Since by Theorem 5.1.2 such a solution is unique and it is given by $P_t^*\mu$, $t \geq 0$, we have $\int_H \varphi(x)P_t^*\mu(dx) = \int_H \varphi(x)\mu_t(dx)$, for any $\varphi \in \mathcal{E}_A(H)$. By taking into account that the set $\mathcal{E}_A(H)$ is π-dense in $C_b(H)$ (cf. Proposition 3.1.3), we have $\int_H \varphi(x)P_t^*\mu(dx) = \int_H \varphi(x)\mu_t(dx)$, for any $\varphi \in C_b(H)$, which implies $P_t^*\mu = \mu_t$, $\forall t \geq 0$. This concludes the proof.

Chapter 6
The reaction-diffusion operator

We deal with Kolmogorov operators associated to reaction-diffusion sto-chastic equations. The results of Theorems 6.2.1, 6.2.2, 6.2.3 seem to be new and they are contained in the submitted paper [38].

6.1. Introduction

We shall consider here the stochastic heat equation perturbed by a poly-nomial term of odd degree $d > 1$ having negative leading coefficient (this will ensures non explosion). We shall represent this polynomial as

$$\lambda \xi - p(\xi), \quad \xi \in \mathbb{R},$$

where $\lambda \in \mathbb{R}$ and p is an increasing polynomial, that is $p'(\xi) \geq 0$ for all $\xi \in \mathbb{R}$.

We set $H = L^2(\mathcal{O})$ where $\mathcal{O} = [0, 1]^n$, $n \in \mathbb{N}$, and denote by $\partial \mathcal{O}$ the boundary of $\mathcal{O}$. We are concerned with the following stochastic differen-tial equation with Dirichlet boundary conditions

$$\begin{cases} dX(t, \xi) = [\Delta_\xi X(t, \xi) + \lambda X(t, \xi) - p(X(t, \xi))]dt + BdW(t, \xi), \\ \qquad \xi \in \mathcal{O}, \\ X(t, \xi) = 0, \quad t \geq 0, \ \xi \in \partial \mathcal{O}, \\ \\ X(0, \xi) = x(\xi), \quad \xi \in \mathcal{O}, \ x \in H, \end{cases}$$

$$(6.1)$$

where Δ_ξ is the Laplace operator, $B \in \mathcal{L}(H)$ and W is a cylindrical Wiener process defined in a stochastic basis $(\Omega, \mathcal{F}, (\mathcal{F}_t)_{t \geq 0}, \mathbb{P})$ in H. We choose W of the form

$$W(t, \xi) = \sum_{k=1}^{\infty} e_k(\xi)\beta_k(t), \quad \xi \in \mathcal{O}, \ t \geq 0,$$

where $\{e_k\}$ is a complete orthonormal system in H and $\{\beta_k\}$ a sequence of mutually independent standard Brownian motions on $(\Omega, \mathcal{F}, (\mathcal{F}_t)_{t\geq 0}, \mathbb{P})$.

Let us write problem (6.1) as a stochastic differential equation in the Hilbert space H. For this we denote by A the realization of the Laplace operator with Dirichlet boundary conditions,

$$\begin{cases} Ax = \Delta_\xi x, & x \in D(A), \\ D(A) = H^2(\mathcal{O}) \cap H_0^1(\mathcal{O}). \end{cases} \tag{6.2}$$

A is self–adjoint and has a complete orthonormal system of eigenfunctions, namely

$$e_k(\xi) = (2/\pi)^{n/2} \sin(\pi k_1 \xi_1) \cdots (\sin \pi k_n \xi_n),$$

where $k = (k_1, ..., k_n)$, $k_i \in \mathbb{N}$. For any $x \in H$ we set $x_k = \langle x, e_k \rangle$, $k \in \mathbb{N}^n$. Notice that

$$Ae_k = -\pi^2 |k|^2, \quad k \in \mathbb{N}^n, \ |k|^2 = k_1^2 + \cdots + k_n^2.$$

Therefore, we have

$$\|e^{tA}\| \leq e^{-\pi^2 t}, \quad t \geq 0. \tag{6.3}$$

Remark 6.1.1. *We can also consider the realization of the Laplace operator Δ with Neumann boundary conditions*

$$\begin{cases} Nx = \Delta_\xi x, & x \in D(N), \\ D(N) = \left\{ x \in H^2(\mathcal{O}) : \dfrac{\partial x}{\partial \eta} = 0 \text{ on } \partial\mathcal{O} \right\} \end{cases}$$

where η represents the outward normal to $\partial\mathcal{O}$. Then

$$Nf_k = -\pi^2 |k|^2 f_k, \quad k \in (\mathbb{N} \cap \{0\})^n,$$

where

$$f_k(\xi) = (2/\pi)^{n/2} \cos(\pi k_1 \xi_1) \cdots (\cos \pi k_n \xi_n),$$

$k = (k_1, ..., k_n)$, $k_i \in \mathbb{N} \cup \{0\}$ *and* $|k|^2 = k_1^2 + \cdots + k_n^2.$

Concerning the operator B we shall assume, for the sake of simplicity,[1] that $B = (-A)^{-\gamma/2}$, where

$$\gamma > \frac{n}{2} - 1. \tag{6.4}$$

The reason of this assumption will be explained in section 6.4.

---

[1] All following results remain true taking $B = G(-A)^{-\gamma/2}$ with $G \in \mathcal{L}(H)$.

Now, setting $X(t) = X(t, \cdot)$ and $W(t) = W(t, \cdot)$, we shall write problem (6.1) as

$$\begin{cases} dX(t) = [AX(t) + F(X(t))]dt + (-A)^{-\gamma/2}dW(t), \\ X(0) = x. \end{cases} \tag{6.5}$$

where F is the mapping

$$F : D(F) = L^{2d}(\mathcal{O}) \subset H \to H, \ x(\xi) \mapsto \lambda\xi - p(x(\xi)).$$

It is well known that for any $x \in L^{2d}(\mathcal{O})$ problem (6.5) has a unique mild solution $X(t, x)$, $t \geq 0$, $x \in H$ (see, for instance, [9, 13]), fulfilling

$$X(t, x) = e^{tA}x + \int_0^t e^{(t-s)A} Bd W(s) + \int_0^t e^{(t-s)A} F(X(s, x))ds \tag{6.6}$$

for any $t \geq 0$. Finally, it is well known that for any $T > 0$ there exists $c > 0$ such that

$$\sup_{t \in [0,T]} \mathbb{E}\left[|X(t, x)|_{L^{2d}(\mathcal{O})}^d\right] \leq c\left(1 + |x|_{L^{2d}(\mathcal{O})}^d\right). \tag{6.7}$$

$$|X(t, x) - X(t, y)| \leq e^{(\lambda - \pi^2)t}|x - y|, \tag{6.8}$$

see [13], Theorem 4.8.

6.2. Main results

We consider here the Kolmogorov operator

$$K_0\varphi(x) = \frac{1}{2}\mathrm{Tr}\left[BB^* D^2\varphi(x)\right] + \langle x, AD\varphi(x)\rangle + \langle D\varphi(x), F(x)\rangle, \ x \in L^{2d}(\mathcal{O}). \tag{6.9}$$

We are interested in extending the results of Theorems 5.1.2, 5.1.3, 5.1.4 to this operator. This will be done in Theorems 6.2.1, 6.2.2, 6.2.3 respectively.

Denote by $C_{b,d}(L^{2d}(\mathcal{O}))$ the space of all functions $\varphi : L^{2d}(\mathcal{O}) \to \mathbb{R}$ such that the function

$$L^{2d}(\mathcal{O}) \to \mathbb{R}, \quad x \to \frac{\varphi(x)}{1 + |x|_{L^{2d}(\mathcal{O})}^d}$$

is uniformly continuous and bounded. The space $C_{b,d}(L^{2d}(\mathcal{O}))$, endowed with the norm

$$\|\varphi\|_{C_{b,d}(L^{2d}(\mathcal{O}))} = \sup_{x \in L^{2d}(\mathcal{O})} \frac{|\varphi(x)|}{1 + |x|_{L^{2d}(\mathcal{O})}^d}$$

is a Banach space. For a sequence $(\varphi_n)_{n\in\mathbb{N}} \subset C_{b,d}(L^{2d}(\mathcal{O}))$ and a function $\varphi \in C_{b,d}(L^{2d}(\mathcal{O}))$ we write

$$\lim_{n\to\infty} \frac{\varphi_n}{1+|\cdot|^d_{L^{2d}(\mathcal{O})}} \stackrel{\pi}{=} \frac{\varphi}{1+|\cdot|^d_{L^{2d}(\mathcal{O})}}.$$

if

$$\lim_{n\to\infty} \frac{\varphi_n(x)}{1+|x|^d_{L^{2d}(\mathcal{O})}} = \frac{\varphi(x)}{1+|x|^d_{L^{2d}(\mathcal{O})}}, \quad \forall x \in L^{2d}(\mathcal{O})$$

and

$$\sup_{n\in\mathbb{N}} \|\varphi_n\|_{C_{b,d}(L^{2d}(\mathcal{O}))} < \infty.$$

Thanks to estimates (6.7) and (6.8) we can define a semigroup of transition operators in $C_{b,d}(L^{2d}(\mathcal{O}))$, by the formula

$$P_t\varphi(x) = \mathbb{E}\big[\varphi(X(t,x))\big], \quad t \geq 0, \ \varphi \in C_{b,d}(L^{2d}(\mathcal{O})), \ x \in L^{2d}(\mathcal{O}),$$
$$(6.10)$$

see Proposition 6.3.1. We define its infinitesimal generator by setting

$$\begin{cases} D(K, C_{b,d}(L^{2d}(\mathcal{O}))) = \Big\{\varphi \in C_{b,d}(L^{2d}(\mathcal{O})) : \exists g \in C_{b,d}(L^{2d}(\mathcal{O})), \\ \qquad \lim_{t\to 0^+} \frac{P_t\varphi(x)-\varphi(x)}{t} = g(x), \forall x \in L^{2d}(\mathcal{O}), \\ \qquad \sup_{t\in(0,1)} \left\|\frac{P_t\varphi-\varphi}{t}\right\|_{C_{b,d}(L^{2d}(\mathcal{O}))} < \infty\Big\} \\ K\varphi(x) = \lim_{t\to 0^+} \frac{P_t\varphi(x)-\varphi(x)}{t}, \quad \varphi \in D(K, C_{b,d}(L^{2d}(\mathcal{O}))), \ x \in L^{2d}(\mathcal{O}). \end{cases}$$
$$(6.11)$$

We recall that $\mathcal{M}_d(L^{2d}(\mathcal{O}))$ is the space of all finite Borel measures on $L^{2d}(\mathcal{O})$ such that

$$\int_{L^{2d}(\mathcal{O})} |x|^d_{L^{2d}(\mathcal{O})} |\mu|_{TV}(dx) < \infty.$$

Since $L^{2d}(\mathcal{O}) \subset H$, we have $\mathcal{M}_d(L^{2d}(\mathcal{O})) \subset \mathcal{M}(H)$. The following theorem generalizes Theorem 5.1.2 to the reaction-diffusion case.

Theorem 6.2.1. *Let $(P_t)_{t\geq 0}$ be the semigroup defined by (6.10) in $C_{b,2}(H)$, and let $(K, D(K, C_{b,d}(L^{2d}(\mathcal{O}))))$ be its infinitesimal generator in $C_{b,d}(L^{2d}(\mathcal{O}))$, defined by (6.11). Then, the formula*

$$\langle\varphi, P_t^* F\rangle_{\mathcal{L}(C_{b,d}(L^{2d}(\mathcal{O})), (C_{b,d}(L^{2d}(\mathcal{O})))^*)} = \langle P_t\varphi, F\rangle_{\mathcal{L}(C_{b,d}(L^{2d}(\mathcal{O})), (C_{b,d}(L^{2d}(\mathcal{O})))^*)}$$

defines a semigroup $(P_t^*)_{t \geq 0}$ of linear and bounded operators on $(C_{b,d}(L^{2d}(\mathcal{O})))^*$ which maps $\mathcal{M}_d(L^{2d}(\mathcal{O}))$ into $\mathcal{M}_d(L^{2d}(\mathcal{O}))$. Moreover, for any $\mu \in \mathcal{M}_d(L^{2d}(\mathcal{O}))$ there exists a unique family of measures $\{\mu_t, \ t \geq 0\} \subset \mathcal{M}_d(L^{2d}(\mathcal{O}))$ such that

$$\int_0^T \left(\int_H |x|^d_{L^{2d}(\mathcal{O})} |\mu_t|_{TV}(dx) \right) dt < \infty, \quad \forall T > 0 \qquad (6.12)$$

and

$$\int_H \varphi(x)\mu_t(dx) - \int_H \varphi(x)\mu(dx) = \int_0^t \left(\int_H K\varphi(x)\mu_s(dx) \right) ds, \quad (6.13)$$

$t \geq 0$, $\varphi \in D(K, C_{b,d}(L^{2d}(\mathcal{O})))$. Finally, the solution of (6.12), (6.13) is given by $P_t^*\mu$, $t \geq 0$.

It is worth noticing that $C_b(H) \subset C_{b,1}(H) \subset C_{b,d}(L^{2d}(\mathcal{O}))$, with continuous embedding. This argument will be used in what follows. Note, also, that for any $\varphi \in C_{b,d}(L^{2d}(\mathcal{O}))$ there exists a sequence $(\varphi_n)_{n \in \mathbb{N}} \subset C_b(H)$ such that

$$\lim_{n \to \infty} \frac{\varphi_n}{1 + |\cdot|^d_{L^{2d}(\mathcal{O})}} \stackrel{\pi}{=} \frac{\varphi}{1 + |\cdot|^d_{L^{2d}(\mathcal{O})}}.$$

The main result of this section is the following

Theorem 6.2.2. The operator $(K, D(K, C_{b,d}(L^{2d}(\mathcal{O}))))$ defined in (6.11) is an extension of K_0, and for any $\varphi \in \mathcal{E}_A(H)$ we have $\varphi \in D(K, C_{b,d}(L^{2d}(\mathcal{O})))$ and $K\varphi = K_0\varphi$. Moreover, the set $\mathcal{E}_A(H)$ is a π-core for $(K, D(K, C_{b,d}(L^{2d}(\mathcal{O}))))$, that is for any $\varphi \in D(K, C_{b,d}(L^{2d}(\mathcal{O})))$ there exist $m \in \mathbb{N}$ and an m-indexed sequence $(\varphi_{n_1,\ldots,n_m})_{n_1 \in \mathbb{N}, \ldots, n_m \in \mathbb{N}} \subset \mathcal{E}_A(H)$ such that

$$\lim_{n_1 \to \infty} \cdots \lim_{n_m \to \infty} \frac{\varphi_{n_1,\ldots,n_m}}{1 + |\cdot|^d_{L^{2d}(\mathcal{O})}} \stackrel{\pi}{=} \frac{\varphi}{1 + |\cdot|^d_{L^{2d}(\mathcal{O})}} \qquad (6.14)$$

and

$$\lim_{n_1 \to \infty} \cdots \lim_{n_m \to \infty} \frac{K_0\varphi_{n_1,\ldots,n_m}}{1 + |\cdot|^d_{L^{2d}(\mathcal{O})}} \stackrel{\pi}{=} \frac{K\varphi}{1 + |\cdot|^d_{L^{2d}(\mathcal{O})}}. \qquad (6.15)$$

Thanks to Theorem 6.2.2 we are able to prove the following

Theorem 6.2.3. For any $\mu \in \mathcal{M}_d(L^{2d}(\mathcal{O}))$ there exists a unique family of measures $\{\mu_t, \ t \geq 0\} \subset \mathcal{M}_d(L^{2d}(\mathcal{O}))$ fulfilling (6.12) and the measure equation

$$\int_H \varphi(x)\mu_t(dx) - \int_H \varphi(x)\mu(dx) = \int_0^t \left(\int_H K_0\varphi(x)\mu_s(dx) \right) ds, \quad (6.16)$$

$t \geq 0$, $\varphi \in \mathcal{E}_A(H)$. Finally, the solution of (6.12), (6.16) is given by $P_t^*\mu$, $t \geq 0$.

In the next section we study the transition semigroup (6.10) and its infinitesimal generator (6.11) in the space $C_{b,d}(L^{2d}(\mathcal{O}))$. In section 6.4 we shall introduce an approximation of problem (6.5) that will be often used in what follows. Finally, in sections 6.5, 6.6, 6.7 we prove Theorems 6.2.1, 6.2.2, 6.2.3, respectively.

6.3. The transition semigroup in $C_{b,d}(L^{2d}(\mathcal{O}))$

The following two propositions may be proved in essentially the same way as Proposition 5.1.6 and Proposition 5.1.8.

Proposition 6.3.1. *Formula* (6.10) *defines a semigroup of operators* $(P_t)_{t\geq 0}$ *in* $C_{b,d}(L^{2d}(\mathcal{O}))$, *and there exists a family of probability measures* $\{\pi_t(x,\cdot),\ t \geq 0,\ x \in L^{2d}(\mathcal{O})\} \subset \mathcal{M}_d(L^{2d}(\mathcal{O}))$ *and two constants* $c_0, \omega_0 > 0$, *such that*

(i) $P_t \in \mathcal{L}(C_{b,d}(L^{2d}(\mathcal{O})))$ *and* $\|P_t\|_{\mathcal{L}(C_{b,d}(L^{2d}(\mathcal{O})))} \leq c_0 e^{\omega_0 t}$;

(ii) $P_t\varphi(x) = \displaystyle\int_H \varphi(y)\pi_t(x,dy)$, *for any* $t \geq 0$, $\varphi \in C_{b,d}(L^{2d}(\mathcal{O}))$, $x \in L^{2d}(\mathcal{O})$;

(iii) *for any* $\varphi \in C_{b,d}(L^{2d}(\mathcal{O}))$, $x \in H$, *the function* $\mathbb{R}^+ \to \mathbb{R}$, $t \mapsto P_t\varphi(x)$ *is continuous.*

(iv) $P_t P_s = P_{t+s}$, *for any* $t, s \geq 0$ *and* $P_0 = I$;

(v) *for any* $\varphi \in C_{b,d}(L^{2d}(\mathcal{O}))$ *and any sequence* $(\varphi_n)_{n\in\mathbb{N}} \subset C_{b,d}(L^{2d}(\mathcal{O}))$ *such that*

$$\lim_{n\to\infty} \frac{\varphi_n}{1+|\cdot|^d_{L^{2d}(\mathcal{O})}} \overset{\pi}{=} \frac{\varphi}{1+|\cdot|^d_{L^{2d}(\mathcal{O})}}$$

we have, for any $t \geq 0$,

$$\lim_{n\to\infty} \frac{P_t\varphi_n}{1+|\cdot|^d_{L^{2d}(\mathcal{O})}} \overset{\pi}{=} \frac{P_t\varphi}{1+|\cdot|^d_{L^{2d}(\mathcal{O})}}.$$

Proposition 6.3.2. *Under the hypothesis of Proposition 6.3.1, let us consider the infinitesimal generator* $(K, D(K, C_{b,d}(L^{2d}(\mathcal{O}))))$ *of* P_t *defined in* (6.11). *Then*

(i) *for any* $\varphi \in D(K, C_{b,d}(L^{2d}(\mathcal{O})))$, *we have* $P_t\varphi \in D(K, C_{b,d}(L^{2d}(\mathcal{O})))$ *and* $K P_t\varphi = P_t K\varphi$, $t \geq 0$;

(ii) *for any* $\varphi \in D(K, C_{b,d}(L^{2d}(\mathcal{O})))$, $x \in L^{2d}(\mathcal{O})$, *the map* $[0, \infty) \to \mathbb{R}$, $t \mapsto P_t\varphi(x)$ *is continuously differentiable and* $(d/dt)P_t\varphi(x) = P_t K\varphi(x)$;

(iii) *given $c_0, \omega_0 > 0$ as in Proposition 6.3.1, for any $\omega > \omega_0$ the linear operator $R(\omega, K)$ on $C_{b,d}(L^{2d}(\mathcal{O}))$ defined by*

$$R(\omega, K)f(x) = \int_0^\infty e^{-\omega t} P_t f(x) dt,$$

$$f \in C_{b,d}(L^{2d}(\mathcal{O})), \quad x \in L^{2d}(\mathcal{O})$$

satisfies, for any $f \in C_{b,1}(H)$

$$R(\omega, K) \in \mathcal{L}(C_{b,d}(L^{2d}(\mathcal{O}))), \quad \|R(\omega, K)\|_{\mathcal{L}(C_{b,d}(L^{2d}(\mathcal{O})))} \le \frac{c_0}{\omega - \omega_0}$$

$$R(\omega, K)f \in D(K, C_{b,d}(L^{2d}(\mathcal{O}))), \quad (\omega I - K)R(\omega, K)f = f.$$

We call $R(\omega, K)$ the resolvent *of K at ω.*

6.4. Some auxiliary results

It is convenient to consider the Ornstein–Uhlenbeck process

$$\begin{cases} dZ(t) = AZ(t)dt + (-A)^{-\gamma/2} dW(t), \\ Z(0) = x, \end{cases}$$

and the corresponding transition semigroup in $C_{b,1}(H)$

$$R_t \varphi(x) = \mathbb{E}[\varphi(Z(t, x))], \quad \varphi \in C_{b,1}(H). \tag{6.17}$$

Notice that thanks to (6.3), (6.4) the operator

$$Q_t x = \int_0^t e^{sA} BB^* e^{sA^*} x \, ds = \int_0^t (-A)^{-\gamma} e^{2tA} x \, dt$$

$$= \frac{1}{2}(-A)^{-(1+\gamma)}(1 - e^{2tA})x, \quad t \ge 0, x \in H,$$

is of trace class. This implies that the Ornstein-Uhlenbeck process $Z(t, x)$ has Gaussian law of mean $e^{tA}x$ and covariance operator Q_t. For the corresponding transition semigroup the representation formula

$$R_t \varphi(x) = \int_H \varphi(e^{tA}x + y) \mathcal{N}_{Q_t}(dy)$$

holds for any $t \ge 0$, $\varphi \in C_{b,1}(H)$, $x \in H$. Notice that we can take $\gamma = 0$ and $B = I$ (white noise) only for $n = 1$. As in section 5.3.1, we denote by $(L, D(L, C_{b,1}(H)))$ the infinitesimal generator of the Ornstein-Uhlenbeck semigroup $(R_t)_{t \ge 0}$ in the space $C_{b,1}(H)$.

A basic tool for proving our results is provided by the following approximating problem

$$\begin{cases} dX^n(t) = (AX^n(t) + F_n(X^n(t)))dt + (-A)^{-\gamma/2}dW(t), \\ X^n(0) = x \in H, \end{cases} \tag{6.18}$$

where for any $n \in \mathbb{N}$, $F_n : H \to H$ is defined by

$$F_n(x)(\xi) = \lambda x(\xi) - p_n(x(\xi)), \quad x \in H$$

and p_n is defined by

$$p_n(\eta) = \frac{np(\eta)}{\sqrt{n^2 + p^2(\eta)}}, \quad \eta \in \mathbb{R}.$$

Notice that p_n is bounded and differentiable, with bounded differential

$$p'_n(\eta) = \frac{np'(\eta)}{\sqrt{n^2 + p^2(\eta)}} \left(1 - \frac{p^2(\eta)}{n^2 + p^2(\eta)} \right) \geq 0,$$

for any $n \in \mathbb{N}$, $\eta \in \mathbb{R}$. Clearly, $|p_n(\eta)| \leq |p(\eta)|$, $\eta \in \mathbb{R}$ and $p_n(\eta) \to p(\eta)$ as $n \to \infty$, for any $\eta \in \mathbb{R}$. The mapping $F_n : H \to H$ is Lipschitz continuous, and for any $n \in \mathbb{N}$, $x \in H$ problem (6.18) has a unique mild solution $X^n(t, x), t \geq 0$ (cf. Section 1). Since by the above discussion we have $|F_n(x)| \leq |F(x)|$, $x \in H$ and $|F_n(x)| \to |F(x)|$ as $n \to \infty$, for any $x \in H$ it is not difficult though tedious to show that for any $x \in L^{2d}(\mathcal{O})$ it holds

$$\lim_{n \to \infty} \sup_{t \in [0,T]} \mathbb{E}\left[|X^n(t, x) - X(t, x)|^2 \right] = 0 \tag{6.19}$$

and

$$\mathbb{E}\left[|X^n(t, x)|^d_{L^{2d}(\mathcal{O})} \right] \leq \mathbb{E}\left[|X(t, x)|^d_{L^{2d}(\mathcal{O})} \right], \quad n \in \mathbb{N}. \tag{6.20}$$

Thanks to (6.19), (6.20) and by the fact that $F_n : H \to H$ is Lipschitz continuous, we can define the transition semigroup associated to the mild solution of (6.18) in both the spaces $C_{b,d}(L^{2d}(\mathcal{O}))$ and $C_{b,1}(H)$.

Proposition 6.4.1. *For any $n \in \mathbb{N}$, let $(P^n_t)_{t \geq 0}$ be the transition semigroup associated to the mild solution of problem (6.18) in the space $C_{b,d}(L^{2d}(\mathcal{O}))$, defined as in (6.10) with $X^n(t, x)$ replacing $X(t, x)$. Then*

(i) *$(P^n_t)_{t \geq 0}$ satisfies statements (i)–(v) of Proposition 6.3.1, and for c_0, ω_0 as in Proposition 6.3.1 we have $\| P^n_t \|_{\mathcal{L}(C_{b,d}(L^{2d}(\mathcal{O})))} \leq c_0 e^{\omega_0 t}$;*

(ii) $(P_t^n)_{t\geq 0}$ is a semigroup of operators in the space $C_{b,1}(H)$. Moreover it satisfies statements (i)–(v) of Proposition 5.1.6. In particular, there exists $c_n, \omega_n > 0$ such that $\|P_t^n\|_{\mathcal{L}(C_{b,1}(H))} \leq c_n e^{\omega_n t}$, for any $t \geq 0$.

Proof. (i) follows by (6.20). (ii) follows since equation (6.18) satisfies Hypothesis 5.1.1. $\qquad\square$

By (ii) of Proposition 6.4.1, we can define, for any $n \in \mathbb{N}$, the infinitesimal generator $(K_n, D(C_{b,1}(H)))$ of the semigroup $(P_t^n)_{t\geq 0}$ in the space $C_{b,1}(H)$ (cf. (5.6)).

By Theorem 5.3.1 and Proposition 6.4.1 it follows that

Proposition 6.4.2. *For any* $n \in \mathbb{N}$ *we have that* $D(L, C_{b,1}(H)) \cap C_b^1(H) = D(K_n, C_{b,1}(H)) \cap C_b^1(H)$, *and for any* $\varphi \in D(L, C_{b,1}(H)) \cap C_b^1(H)$ *we have* $K_n\varphi = L\varphi + \langle D\varphi, F_n\rangle$.

The semigroup $(P_t^n)_{t\geq 0}$ enjoys the following property, which will be essential in the proof of Theorem 6.2.2.

Proposition 6.4.3. *For any* $n \in \mathbb{N}$, *the semigroup* $(P_t^n)_{t\geq 0}$ *maps* $C_b^1(H)$ *into* $C_b^1(H)$, *and for any* $\varphi \in C_b^1(H)$ *it holds*

$$|DP_t\varphi(x)| \leq e^{2(\lambda - \pi^2)t} \sup_{x\in H} |D\varphi(x)|.$$

Proof. Since the nonlinear mapping F_n is differentiable with uniformly continuous and bounded differential, it is well known (see, for instance, [23]) that the mild solution $X^n(t, x)$ of problem (6.18) is differentiable with respect to x and for any $x, h \in H$ we have $DX^n(t, x) \cdot h = \eta_n^h(t, x)$, where $\eta_n^h(t, x)$ is the mild solution of the differential equation with random coefficients

$$\begin{cases} \dfrac{d}{dt}\eta_n^h(t, x) = A\eta_n^h(t, x) + DF_n(X^n(t, x)) \cdot \eta_n^h(t, x) & t \geq 0 \\ \eta_n^h(t, x) = 0. \end{cases}$$

Multiplying the above identity by $\eta_n^h(t, x)$ and integrating over $\mathcal{O}$ we find that

$$\frac{1}{2}\frac{d}{dt}|\eta_n^h(t, x)|^2 = \langle (A + \lambda)\eta_n^h(t, x), \eta_n^h(t, x)\rangle$$
$$- \int_{\mathcal{O}} p_n'(X^n(t, x)(\xi))|\eta_n^h(t, x)(\xi)|^2 d\xi.$$

Taking into account that $p_n' \geq 0$ and integrating by parts we find that

$$\frac{1}{2}\frac{d}{dt}|\eta_n^h(t, x)|^2 + \int_{\mathcal{O}} |D_\xi\eta_n^h(t, x)(\xi)|^2 d\xi \leq \lambda|\eta_n^h(t, x)|^2.$$

Now, the classical Poincaré inequality implies $|D_\xi \eta_n^h(t,x)| \geq \pi^2 |\eta_n^h(t,x)|$ and so we obtain

$$\frac{1}{2}\frac{d}{dt}|\eta_n^h(t,x)|^2 \leq (\lambda - \pi^2)|\eta_n^h(t,x)|^2, \quad x \in H, t \geq 0.$$

Consequently, by the Gronwall lemma we find that

$$|\eta^h(t,x)| \leq e^{2(\lambda-\pi^2)t}|h|. \tag{6.21}$$

Now take $\varphi \in C_b^1(H)$. For any $x, h \in H$ we have

$$DP_t^n\varphi(x) \cdot h = \mathbb{E}\left[D\varphi(X^n(t,x)) \cdot \eta^h(t,x)\right].$$

Hence by (6.21)

$$|DP_t^n\varphi(x) \cdot h| \leq \mathbb{E}\left[|D\varphi(X^n(t,x))||\eta^h(t,x)|\right] \leq \sup_{x\in H}|D\varphi(x)|e^{2(\lambda-\pi^2)t}|h|,$$

which implies the result. $\qquad\square$

6.5. Proof of Theorem 6.2.1

We have first to show that $(P_t^*)_{t\geq0}$ is a semigroup of linear and continuous operators in $(C_{b,d}(L^{2d}(\mathcal{O})))^*$ and that $P_t^*\mu \in \mathcal{M}_d(L^{2d}(\mathcal{O}))$ for any $t \geq 0$, $\mu \in \mathcal{M}_d(L^{2d}(\mathcal{O}))$. These facts follow by Proposition 6.3.1 and by the argument of Lemma 5.2.2. We leave the details to the reader.

We now show existence of a solution for the measure equation, namely we show that $P_t^*\mu$, $t \geq 0$ fulfils (6.16), (6.12). To show that $P_t^*\mu$, $t \geq 0$ fulfils (6.16) one can use the argument from Lemma 5.2.3. We left the details to the reader. We now check that (6.12) holds. Fix $T > 0$. By the local boundedness of the operators $P_t^*\mu$ and by the semigroup property it follows that there exists $c > 0$ such that

$$\sup_{t\in[0,T]} \|P_t^*\|_{\mathcal{L}((C_{b,d}(L^{2d}(\mathcal{O})))^*)} \leq c.$$

Still by the first part of the theorem, since $\mu \in \mathcal{M}_d(L^{2d}(\mathcal{O}))$ we have $P_t^*\mu \in \mathcal{M}_d(L^{2d}(\mathcal{O}))$. Hence

$$\int_0^T \left(\int_H |x|_{L^{2d}(\mathcal{O})}^d |P_t^*\mu|_{TV}(dx)\right) dt$$

$$= \int_0^T \left(\int_{L^{2d}(\mathcal{O})} |x|_{L^{2d}(\mathcal{O})}^d |P_t^*\mu|_{TV}(dx)\right) dt$$

$$\leq \int_0^T \|P_t^*\mu\|_{(C_{b,d}(L^{2d}(\mathcal{O})))^*} dt \leq c\int_0^T \|\mu\|_{(C_{b,d}(L^{2d}(\mathcal{O})))^*} dt$$

$$= cT\|\mu\|_{(C_{b,d}(L^{2d}(\mathcal{O})))^*} = cT \int_{L^{2d}(\mathcal{O})} (1 + |x|_{L^{2d}(\mathcal{O})}^d)|\mu|_{TV}(dx) < \infty.$$

Then, (6.12) is proved.

Let us prove uniqueness of the solution. By (5.3) follows that the mild solution $X(t, x)$ of problem (6.6) can be extended to a process $(X(t, x))_{t \geq 0, x \in H}$ with values in H and adapted to the filtration $(\mathcal{F}_t)_{t \geq 0}$. In the literature, the process $X(\cdot, x)$ is called a *generalized solution* of equation (6.6) (see [13]). Hence, we can extend the transition semigroup (6.10) to a semigroup in $C_b(H)$, still denoted by $(P_t)_{t \geq 0}$, by setting

$$P_t \varphi(x) = \mathbb{E}\left[\varphi(X(t, x))\right] \quad t \geq 0, \ x \in H, \ \varphi \in C_b(H).$$

Clearly, $\|P_t\|_{\mathcal{L}(C_b(H))} \leq 1$. In addition, the representation

$$P_t \varphi(x) = \int_H \varphi(y) \pi_t'(x, dy)$$

holds for any $\varphi \in C_b(H)$, where $\pi_t'(x, \cdot)$ is the probability measure on H defined by $\pi_t'(x, \Gamma) = \mathbb{P}(X(t, x) \in \Gamma)$, $\Gamma \in \mathcal{B}(H)$. It is clear that $\pi_t'(x, \Gamma) = \pi_t(x, \Gamma)$ when $\Gamma \in \mathcal{B}(L^{2d}(\mathcal{O}))$. We define the infinitesimal generator $K : D(K, C_b(H)) \to C_b(H)$ of the semigroup $(P_t)_{t \geq 0}$ in the space $C_b(H)$ as in (5.13). By arguing as in Lemma 5.2.3, one can show that the semigroup $(P_t)_{t \geq 0}$ in $C_b(H)$ is a stochastically continuous Markov semigroup, in the sense of [37]. So, we can apply Theorem 5.2.1 and then for any $\mu \in \mathcal{M}(H)$ there exists a unique family of measures $\{\mu_t, \ t \geq 0\} \subset \mathcal{M}(H)$ such that

$$\int_0^T |\mu_t|_{TV}(H) dt < \infty, \quad \forall T > 0 \tag{6.22}$$

and (6.13) holds for any $t \geq 0$ and $\varphi \in D(K, C_b(H))$.

Now take $\mu = 0$, and assume that $\{\mu_t, \ t \geq 0\} \subset \mathcal{M}_d(L^{2d}(\mathcal{O}))$ fulfils (6.16), (6.12). Then $\{\mu_t, \ t \geq 0\} \subset \mathcal{M}(H)$. We want to show now that $\mu_t, t \geq 0$ fulfils also (6.22) and (6.13) for any $t \geq 0$, $\varphi \in D(K, C_b(H))$. Taking in mind that for this equation the solution is unique, this will imply $\mu_t = 0$ (as measure in H and consequently as measure in $L^{2d}(\mathcal{O})$) for any $t \geq 0$.

Clearly, (6.22) follows by (6.16). It is also possible to prove, by a standard argument, that $D(K, C_b(H)) \subset D(K, C_{b,d}(L^{2d}(\mathcal{O})))$ and $D(K, C_b(H)) = \{\varphi \in D(K, C_{b,d}(L^{2d}(\mathcal{O}))) \cap C_b(H) : K\varphi \in C_b(H)\}$. Then, for any $\varphi \in D(K, C_b(H))$, we have $\varphi \in D(K, C_{b,d}(L^{2d}(\mathcal{O})))$ and hence (6.13) holds for any $\varphi \in D(K, C_b(H))$. This concludes the proof. $\square$

6.6. Proof of Theorem 6.2.2

The proof is splitted into two lemmata.

Lemma 6.6.1. $(K, D(K, C_{b,d}(L^{2d}(\mathcal{O}))))$ *is an extension of* K_0, *and for any* $\varphi \in \mathcal{E}_A(H)$ *we have* $\varphi \in D(K, C_{b,d}(L^{2d}(\mathcal{O})))$ *and* $K\varphi = K_0\varphi$.

Proof. It is sufficient to prove the claim for $\varphi \in \mathcal{E}_A(H)$ of the form $\varphi(x) = e^{i\langle x,h \rangle}$, $x \in H$, where $h \in D(A)$. For any $t \geq 0$, $x \in L^{2d}(\mathcal{O})$ we have

$$X(t, x) = Z(t, x) + \int_0^t e^{(t-s)A} F(X(s, x))ds,$$

where $Z(t, x)$ is the OU process introduced in section 6.4. Then we have

$$R_t\varphi(x) - \varphi(x) = P_t\varphi(x) - \varphi(x)$$

$$-i\mathbb{E}\left[\int_0^1 \varphi(\xi Z(t, x) + (1-\xi)X(t, x))\left\langle h, \int_0^t e^{(t-s)A} F(X(s, x))ds \right\rangle d\xi \right],$$

for any $x \in L^{2d}(\mathcal{O})$, since $\langle D\varphi(x), y \rangle = i\varphi(x)\langle h, y \rangle$. Since $Z(t, x)$, $X(t, x)$ are continuous in mean square, by arguing as for Theorem 5.3.1 it follows

$$\lim_{t \to 0^+} i\mathbb{E}\left[\frac{1}{t}\int_0^1 \varphi(\xi Z(t, x) \right.$$

$$\left. +(1 - \xi)X(t, x))\left\langle h, \int_0^t e^{(t-s)A} F(X(s, x))ds \right\rangle d\xi \right] \quad (6.23)$$

$$= i\varphi(x)\langle h, F(x) \rangle = \langle D\varphi(x), F(x) \rangle.$$

In addition, by (6.7) we have that for any $x \in L^{2d}(\mathcal{O})$

$$\frac{1}{t}\left|\mathbb{E}\left[\int_0^1 \varphi(\xi Z(t, x)+(1-\xi)X(t, x))\left\langle h, \int_0^t e^{(t-s)A} F(X(s, x))ds \right\rangle d\xi \right]\right|$$

$$\leq \frac{|h|}{t}\mathbb{E}\left[\left|\int_0^t e^{(t-s)A} F(X(s, x))ds \right|\right] \leq \frac{|h|}{t}\int_0^t \mathbb{E}\left[|F(X(s, x))|\right]ds$$

$$\leq \frac{|h|}{t}\int_0^t \mathbb{E}\left[|X(s, x)|_{L^{2d}(\mathcal{O})}^d ds\right] \leq |h|c\left(1 + |x|_{L^{2d}(\mathcal{O})}^d\right).$$

$$(6.24)$$

for some $c > 0$.

We recall that $\mathcal{E}_A(H) \subset D(L, C_{b,1}(H)) \cap C_b^1(H)$. by (6.23) and Proposition 5.3.2 it follows

$$\lim_{t \to 0^+} \frac{P_t\varphi(x) - \varphi(x)}{t} = L\varphi(x) + \langle D\varphi(x), F(x) \rangle = K_0\varphi(x), \quad \forall x \in L^{2d}(\mathcal{O}).$$

By (6.24) we have

$$\sup_{t \in (0,1]} \left\| \frac{P_t\varphi - \varphi}{t} \right\|_{C_{b,d}(L^{2d}(\mathcal{O}))} \leq \sup_{t \in (0,1)} \left\| \frac{R_t\varphi - \varphi}{t} \right\|_{0,1} + |h|c < \infty$$

that implies $\varphi \in D(K, C_{b,d}(L^{2d}(\mathcal{O})))$. This concludes the proof. $\square$

Lemma 6.6.2. *The set $\mathcal{E}_A(H)$ is a π-core for $(K, D(K, C_{b,d}(L^{2d}(\mathcal{O}))))$, and for any $\varphi \in D(K, C_{b,d}(L^{2d}(\mathcal{O})))$ there exists $m \in \mathbb{N}$ and an m-indexed sequence $(\varphi_{n_1,\dots,n_m}) \subset \mathcal{E}_A(H)$ such that*

$$\lim_{n_1 \to \infty} \cdots \lim_{n_m \to \infty} \frac{\varphi_{n_1,\dots,n_m}}{1 + |\cdot|^d_{L^{2d}(\mathcal{O})}} \overset{\pi}{=} \frac{\varphi}{1 + |\cdot|^d_{L^{2d}(\mathcal{O})}}, \qquad (6.25)$$

$$\lim_{n_1 \to \infty} \cdots \lim_{n_m \to \infty} \frac{K_0\varphi_{n_1,\dots,n_m}}{1 + |\cdot|^d_{L^{2d}(\mathcal{O})}} \overset{\pi}{=} \frac{K\varphi}{1 + |\cdot|^d_{L^{2d}(\mathcal{O})}}. \qquad (6.26)$$

Proof. Take $\varphi \in D(K, C_{b,d}(L^{2d}(\mathcal{O})))$. We shall construct the claimed sequence in four steps.

Step 1. Fix $\omega > \omega_0, 2(\lambda - \pi^2)$ and set $f = \omega\varphi - K\varphi$. Then we have $\varphi = R(\omega, K)f$. We approximate f as follows: for any $n_1 \in \mathbb{N}$ we set

$$f_{n_1}(x) = \frac{n_1 f(e^{\frac{1}{n_1}A}x)}{n_1 + |e^{\frac{1}{n_1}A}x|^d_{L^{2d}(\mathcal{O})}}, \qquad x \in H.$$

By the well known properties of the heat semigroup, we have $e^{\frac{1}{n_1}A}x \in L^{2d}(\mathcal{O})$, for any $x \in H$. Hence, $f_{n_1} \in C_b(H)$ and

$$\lim_{n_1 \to \infty} \frac{f_{n_1}}{1 + |\cdot|^d_{L^{2d}(\mathcal{O})}} \overset{\pi}{=} \frac{f}{1 + |\cdot|^d_{L^{2d}(\mathcal{O})}}.$$

By Proposition 6.3.1 we have

$$\lim_{n_1 \to \infty} \frac{P_t f_{n_1}}{1 + |\cdot|^d_{L^{2d}(\mathcal{O})}} \overset{\pi}{=} \frac{P_t f}{1 + |\cdot|^d_{L^{2d}(\mathcal{O})}}$$

for any $t \geq 0$. Since we have $\|P_t\|_{\mathcal{L}(C_{b,d}(L^{2d}(\mathcal{O})))} \leq c_0 e^{\omega_0 t}, \forall t \geq 0$ (*cf.* (i) of Proposition 6.3.1) and $\omega > \omega_0$, it follows

$$\lim_{n_1 \to \infty} \frac{R(\omega, K)f_{n_1}}{1 + |\cdot|^d_{L^{2d}(\mathcal{O})}} \overset{\pi}{=} \frac{R(\omega, K)f}{1 + |\cdot|^d_{L^{2d}(\mathcal{O})}}.$$

Setting $\varphi_{n_1} = R(\omega, K)f_{n_1}$, by the above argument we have

$$\lim_{n_1 \to \infty} \frac{\varphi_{n_1}}{1 + |\cdot|^d_{L^{2d}(\mathcal{O})}} \overset{\pi}{=} \frac{\varphi}{1 + |\cdot|^d_{L^{2d}(\mathcal{O})}},$$

$$\lim_{n_1 \to \infty} \frac{K\varphi_{n_1}}{1 + |\cdot|^d_{L^{2d}(\mathcal{O})}} \overset{\pi}{=} \frac{K\varphi}{1 + |\cdot|^d_{L^{2d}(\mathcal{O})}}. \qquad (6.27)$$

Step 2. For any $n_1 \in \mathbb{N}$, let us fix a sequence $(f_{n_1,n_2})_{n_2 \in \mathbb{N}} \subset C_b^1(H)$ such that

$$\lim_{n_2 \to \infty} f_{n_1,n_2} \overset{\pi}{=} f_{n_1}.$$

Now set $\varphi_{n_1,n_2} = R(\omega, K)f_{n_1,n_2}$. By arguing as in step 1 we have

$$\lim_{n_2 \to \infty} \varphi_{n_1,n_2} \overset{\pi}{=} \varphi_{n_1}, \qquad \lim_{n_2 \to \infty} K\varphi_{n_1,n_2} \overset{\pi}{=} K\varphi_{n_1}. \tag{6.28}$$

Step 3. We now consider the approximation of K introduced in section 6.4. We denote by $(K_{n_3}, D(K_{n_3}, C_{b,1}(H)))$ the infinitesimal generator of the transition semigroup associated to the mild solution of problem (6.18) in the space $C_{b,1}(H)$. For any $n_1, n_2, n_3 \in \mathbb{N}$ set

$$\varphi_{n_1,n_2,n_3} = \int_0^\infty e^{-\omega t} P_t^{n_3} f_{n_1,n_2} dt.$$

For any $n_1, n_2, n_3 \in \mathbb{N}$ the function φ_{n_1,n_2,n_3} is bounded, since

$$\left| \int_0^\infty e^{-\omega t} P_t^{n_3} f_{n_1,n_2} dt \right| \le \|f\|_0 \int_0^\infty e^{-\omega t} dt < \infty.$$

The fact that $\varphi_{n_1,n_2,n_3} \in C_b(H)$ follows by standard computations. By (v) of Proposition 5.1.6 and by (i) of Proposition 6.4.1 it follows that

$$\lim_{n_3 \to \infty} \frac{\varphi_{n_1,n_2,n_3}}{1 + |\cdot|_{L^{2d}(\mathcal{O})}^d} \overset{\pi}{=} \frac{\varphi_{n_1,n_2}}{1 + |\cdot|_{L^{2d}(\mathcal{O})}^d}, \tag{6.29}$$

It is also stardard to show that for any $n_1, n_2, n_3 \in \mathbb{N}$ it holds $\varphi_{n_1,n_2,n_3} \in D(K_{n_3}, C_{b,1}(H)))$ for and $K_{n_3}\varphi_{n_1,n_2,n_3} = \omega\varphi_{n_1,n_2,n_3} - f_{n_1,n_2}$. Hence, by (6.29) we obtain

$$\lim_{n_3 \to \infty} \frac{K_{n_3}\varphi_{n_1,n_2,n_3}}{1 + |\cdot|_{L^{2d}(\mathcal{O})}^d} \overset{\pi}{=} \frac{K\varphi_{n_1,n_2}}{1 + |\cdot|_{L^{2d}(\mathcal{O})}^d}. \tag{6.30}$$

By Proposition 6.4.3 it follows that $\varphi_{n_1,n_2,n_3} \in C_b^1(H)$ and

$$|D\varphi_{n_1,n_2,n_3}(x)| = \left| \int_0^\infty e^{-\omega t} DP_t^{n_3} f_{n_1,n_2}(x) dt \right|$$

$$\le \int_0^\infty e^{-(\omega - 2\lambda + 2\pi^2)t} dt \sup_{x \in H} |Df_{n_1,n_2}(x)| \le \frac{\sup_{x \in H} |Df_{n_1,n_2}(x)|}{\omega - 2(\lambda - \pi^2)}. \tag{6.31}$$

Hence $\varphi_{n_1,n_2,n_3} \in D(K_{n_3}, C_{b,1}(H))) \cap C_b^1(H)$, and by Proposition 6.4.2 it follows that $K_{n_3}\varphi_{n_1,n_2,n_3} = L\varphi_{n_1,n_2,n_3} + \langle D\varphi_{n_1,n_2,n_3}, F_{n_3}\rangle$. Hence, by Lemma 6.6.1 we have, for any $x \in L^{2d}(\mathcal{O})$

$$
\begin{aligned}
K\varphi_{n_1,n_2,n_3}(x) &= L\varphi_{n_1,n_2,n_3}(x) + \langle D\varphi_{n_1,n_2,n_3}(x), F(x)\rangle \\
&= K_{n_3}\varphi_{n_1,n_2,n_3}(x) + \langle D\varphi_{n_1,n_2,n_3}(x), F(x) - F_{n_3}(x)\rangle.
\end{aligned}
\tag{6.32}
$$

We recall that $|F_{n_3}(x)| \le |F(x)| \le c|x|_{L^{2d}(\mathcal{O})}^d$, for any $n_3 \in \mathbb{N}$, $x \in L^{2d}(\mathcal{O})$ and for some $c > 0$. In addition, $|F_{n_3}(x) - F(x)| \to 0$ as $n_3 \to \infty$, for any $x \in L^{2d}(\mathcal{O})$. Consequently, by (6.31) it follows that

$$
\lim_{n_3 \to \infty} \frac{\langle D\varphi_{n_1,n_2,n_3}, F - F_{n_3}\rangle}{1 + |\cdot|_{L^{2d}(\mathcal{O})}^d} \overset{\pi}{=} 0.
\tag{6.33}
$$

Step 4. By Propositon 5.3.2 for any $n_1, n_2, n_3 \in \mathbb{N}$ there exists a sequence[2] $(\varphi_{n_1,n_2,n_3,n_4}) \subset \mathcal{E}_A(H)$ such that

$$
\lim_{n_4 \to \infty} \varphi_{n_1,n_2,n_3,n_4} \overset{\pi}{=} \varphi_{n_1,n_2,n_3},
\tag{6.34}
$$

$$
\lim_{n_4 \to \infty} \frac{\frac{1}{2}\mathrm{Tr}\left[BB^* D^2\varphi_{n_1,n_2,n_3,n_4}\right] + \langle x, AD\varphi_{n_1,n_2,n_3,n_4}\rangle}{1 + |\cdot|} \overset{\pi}{=} \frac{L\varphi_{n_1,n_2,n_3}}{1 + |\cdot|}
\tag{6.35}
$$

and for any $h \in H$

$$
\lim_{n_4 \to \infty} \langle D\varphi_{n_1,n_2,n_3,n_4}, h\rangle \overset{\pi}{=} \langle D\varphi_{n_1,n_2,n_3}, h\rangle.
$$

This, together with the above approximation, implies that for any $n_1, n_2, n_3 \in \mathbb{N}$ we have

$$
\lim_{n_4 \to \infty} \frac{\langle D\varphi_{n_1,n_2,n_3,n_4}, F - F_{n_3}\rangle}{1 + |\cdot|_{L^{2d}(\mathcal{O})}^d} \overset{\pi}{=} \frac{\langle D\varphi_{n_1,n_2,n_3}, F - F_{n_3}\rangle}{1 + |\cdot|_{L^{2d}(\mathcal{O})}^d}.
\tag{6.36}
$$

Step 5. By (6.27), (6.28), (6.29), (6.34) we have

$$
\lim_{n_1 \to \infty} \lim_{n_2 \to \infty} \lim_{n_3 \to \infty} \lim_{n_4 \to \infty} \frac{\varphi_{n_1,n_2,n_3,n_4}}{1 + |\cdot|_{L^{2d}(\mathcal{O})}^d} \overset{\pi}{=} \frac{\varphi}{1 + |\cdot|_{L^{2d}(\mathcal{O})}^d},
$$

and consequently (6.25) follows. We now check

$$
\lim_{n_1 \to \infty} \lim_{n_2 \to \infty} \lim_{n_3 \to \infty} \lim_{n_4 \to \infty} \frac{K_0\varphi_{n_1,n_2,n_3,n_4}}{1 + |\cdot|_{L^{2d}(\mathcal{O})}^d} \overset{\pi}{=} \frac{K\varphi}{1 + |\cdot|_{L^{2d}(\mathcal{O})}^d}.
$$

---

[2] We assume that it has one index.

This will prove (6.26). By Lemma 6.6.1, for any $n_1, n_2, n_3, n_4 \in \mathbb{N}$ we have $K\varphi_{n_1,n_2,n_3,n_4} = K_0\varphi_{n_1,n_2,n_3,n_4}$. Moreover, by Theorem 5.1.3 we have $\varphi_{n_1,n_2,n_3,n_4} \in D(K_{n_3}, C_{b,1}(H)))$ and by (6.32)

$$K_0\varphi_{n_1,n_2,n_3,n_4}(x) = K_{n_3}\varphi_{n_1,n_2,n_3,n_4}(x) + \langle D\varphi_{n_1,n_2,n_3,n_4}(x), F(x) - F_{n_3}(x) \rangle,$$

for any $n_1, n_2, n_3, n_4 \in \mathbb{N}$, $x \in L^{2d}(\mathcal{O})$. By (6.32), (6.35), (6.36) it holds

$$\lim_{n_4 \to \infty} \frac{K_0\varphi_{n_1,n_2,n_3,n_4}}{1 + |\cdot|_{L^{2d}(\mathcal{O})}^d} \overset{\pi}{=} \frac{K_{n_3}\varphi_{n_1,n_2,n_3} + \langle D\varphi_{n_1,n_2,n_3}, F - F_{n_3}\rangle}{1 + |\cdot|_{L^{2d}(\mathcal{O})}^d}.$$

By (6.30), (6.33) it holds

$$\lim_{n_3 \to \infty} \frac{K_{n_3}\varphi_{n_1,n_2,n_3} + \langle D\varphi_{n_1,n_2,n_3}, F - F_{n_3}\rangle}{1 + |\cdot|_{L^{2d}(\mathcal{O})}^d} \overset{\pi}{=} \frac{K\varphi_{n_1,n_2}}{1 + |\cdot|_{L^{2d}(\mathcal{O})}^d}.$$

By (6.27), (6.28) it holds

$$\lim_{n_1 \to \infty}\lim_{n_2 \to \infty} \frac{K\varphi_{n_1,n_2}}{1 + |\cdot|_{L^{2d}(\mathcal{O})}^d} \overset{\pi}{=} \frac{K\varphi}{1 + |\cdot|_{L^{2d}(\mathcal{O})}^d}. \qquad \square$$

6.7. Proof of Theorem 6.2.3

Take $\mu \in \mathcal{M}_d(L^{2d}(\mathcal{O}))$. The fact that $P_t^*\mu, t \geq 0$ fulfils (6.12) and (6.16) follows by Theorems 6.2.1, 6.2.2 and by the fact that $KP_t\varphi = P_tK\varphi = P_tK_0\varphi$, for any $\varphi \in \mathcal{E}_A(H)$ (cf. Proposition 6.3.2 and Lemma 6.6.1). Hence, existence of a solution is proved. Let us show that such a solution is unique. Assume that $\{\mu_t, \ t \geq 0\} \subset \mathcal{M}_d(L^{2d}(\mathcal{O}))$ fulfils (6.12) and (6.16). By Theorem 6.2.2 for any $\varphi \in D(K, C_{b,d}(L^{2d}(\mathcal{O})))$ there exist $m \in \mathbb{N}$ and an m-indexed sequence $(\varphi_{n_1,\dots,n_m})_{n_1 \in \mathbb{N},\dots,n_m \in \mathbb{N}} \subset \mathcal{E}_A(H)$ such that (6.14), (6.15) hold. This, together with (6.12), implies that $\mu_t, \ t \geq 0$ fulfils (6.13) for any $t \geq 0$, $\varphi \in D(K, C_{b,d}(L^{2d}(\mathcal{O})))$ (here we can use the same argument used to prove Theorem 5.1.4). Since the solution of (6.12), (6.13) is unique and it is given by $P_t^*\mu, \ t \geq 0$, it follows $\int_H \varphi(x)P_t^*\mu(dx) = \int_H \varphi(x)\mu_t(dx)$, for any $\varphi \in \mathcal{E}_A(H)$. Hence, since $\mathcal{E}_A(H)$ is π-dense in $C_b(H)$, it follows $\int_H \varphi(x)P_t^*\mu(dx) = \int_H \varphi(x)\mu_t(dx)$, for any $\varphi \in C_b(H)$, that implies $P_t^*\mu = \mu_t, \forall t \geq 0$. This concludes the proof. $\qquad \square$

Chapter 7
The Burgers equation

We consider the *Burgers* equation with Dirichlet boundary conditions perturbed by a white noise. Existence and uniqueness of a mild solution are discussed in [20]. In [19] is proved uniqueness of an invariant measure ν (its existence is proved in [20]). Moreover, still in [19], several estimates are proved in order to ensure that the operator K_0(see (7.10) below) is m-dissipative in $L^2(H; \nu)$. Thanks to these estimates, we are able to prove new results which are described by Theorems 7.2.2, 7.2.3 and are the object of a forthcoming paper.

We recall that in [41, 42] it has been considered a *generalized Burgers* stochastic equation and the associated Kolmogorov operator it has been studied in spaces of continuous functions (see the Introduction of this thesis for a more detailed description). However, in [41, 42] the noise is driven by a trace class operator, whereas in our case the perturbation is a white noise.

7.1. Introduction and preliminaries

We consider the stochastic Burgers equation in the interval $[0, 1]$ with Dirichlet boundary conditions perturbed by a space-time white noise

$$
\begin{cases}
dX = \left(D_\xi^2 X + \frac{1}{2} D_\xi(X^2) \right) dt + dW, & \xi \in [0, 1], t \geq 0, \\
X(t, 0) = X(t, 1) = 0 \\
X(0, \xi) = x(\xi), & \xi \in [0, 1],
\end{cases}
\tag{7.1}
$$

where $x \in L^2(0, 1)$ and W is a cylindrical Wiener process defined in a probability space $(\Omega, \mathcal{F}, \mathbb{P})$ and with values in $L^2(0, 1)$.

Let us write problem (7.1) in an abstract form. We denote by $L^p(0, 1)$, $p \geq 1$, the space of all real valued Lebesque measurable functions x :

$[0, 1] \to \mathbb{R}$ such that

$$|x|_p := \left(\int_0^1 |x(\xi)|^p d\xi \right)^{1/p} < +\infty,$$

and by $L^\infty(0, 1)$ the space of all real valued Lebesque measurable essentially bounded functions endowed with the norm

$$|x|_\infty := \sup_{\xi \in [0,1]} |x(\xi)|.$$

We denote by H the Hilbert space of all Lebesque square integrable function $x : [0, 1] \to \mathbb{R}$, endowed with the norm

$$|x|_2 = \left(\int_0^1 |x(\xi)|^2 d\xi \right)^{\frac{1}{2}}$$

and the inner product

$$\langle x, y \rangle = \int_0^1 x(\xi)y(\xi)d\xi, \quad x, y \in H.$$

As usual, $H^k(0, 1)$, $k \in \mathbb{N}$, is the Sobolev space of all functions in H whose differentials belong to H up to the order k, and $H_0^1(0, 1)$ is the subspace of H^1 of all functions whose trace at 0 and 1 vanishes. We define the unbounded self-adjoint operator A in H by

$$Ax = \frac{\partial^2}{\partial \xi^2} x$$

for x in the domain

$$D(A) = H^2(0, 1) \cap H_0^1(0, 1).$$

and by e^{tA}, $t \geq 0$ the semigroup in H generated by A. Finally, we denote by $\{e_k\}_{k \in \mathbb{N}}$ the orthonormal system in H given by the eigenvectors of A

$$e_k(\xi) = \sqrt{\frac{2}{\pi}} \sin(k\xi), \quad \xi \in [0, 1], \ k \in \mathbb{N}.$$

We have

$$Ae_k = -k^2 e_k, \quad k \in \mathbb{N}.$$

We set

$$b(x) = \frac{1}{2} D_\xi(x^2), \quad x \in D(b) = H_0^1.$$

The operator b enjoys the foundamental property

$$\langle b(x), x \rangle = 0 \quad \text{for all } x \in H_0^1.$$

Thanks to the introduced notations, we write problem (7.1) in the abstract form

$$\begin{cases} dX = (AX + b(X))dt + dW(t), \\ X(0) = x, \quad x \in H. \end{cases} \tag{7.2}$$

As usual, the cylindrical Wiener process $W(t)$ is given (formally) by

$$W(t) = \sum_{k=1}^{\infty} \beta_k(t)e_k, \quad t \geq 0,$$

where $\{\beta_k\}$ is a sequence of mutually independent standard Brownian motions on a stochastic basis $(\Omega, \mathcal{F}, (\mathcal{F}_t)_{t \geq 0}, \mathbb{P})$. We recall that the solution of the linear stochastic equation

$$\begin{cases} dZ(t, x) = AZ(t, x)dt + dW(t), \quad t \geq 0 \\ Z(0, x) = x \in H \end{cases} \tag{7.3}$$

is given by the stochastic convolution

$$Z(t, x) = e^{tA}x + W_A(t), \tag{7.4}$$

see Chapter 3. The process $Z(t, x)$ has a version which is, a.s. for $\omega \in \Omega$, α-Hölder continuous with respect to (t, x), for any $\alpha \in (0, \frac{1}{4})$ (see [23], Theorem 5.20 and Example 5.21). Now set

$$Y(t, x) = X(t, x) - W_A(t).$$

We write (7.2) as

$$\begin{cases} Y(t, x) = e^{tA}x + \int_0^t e^{(t-s)A} \frac{\partial}{\partial \xi} (Y(s, x) + W_A(t, x))^2 \, ds, \\ Y(0, x) = x, \quad x \in H. \end{cases} \tag{7.5}$$

As we shall see, if $z(t) \in L^p(0, 1)$ a.s., then $e^{tA} \frac{\partial}{\partial \xi} z^2 \in L^p(0, 1)$ is bounded. Then the above integral converges and the equation is meaningful.

We say that $X(t, x)$ is a *mild* solution of (7.2) if $Y(t, x) = X(t, x) - W_A(t)$ satisfies (7.5) for a.s. all $\omega \in \Omega$.

The following result is proved in [20].

Theorem 7.1.1. *Let $x \in L^p(0,1)$, $p \geq 2$. Then there exists a unique mild solution of equation (7.2), which belongs a.s. to $C([0,T]; L^p(0,1))$, for any $T > 0$.*

The following estimate is proved in [19].

Proposition 7.1.2. *For any $p \geq 2$, $k \geq 1$, $T > 0$ there exists a constant $c_{p,k,T}$ such that*

$$\mathbb{E}\left[\sup_{t \in [0,T]} |X(t,x)|_p^k\right] \leq c_{p,k,T}(1 + |x|_p^k).$$

7.2. Main results

In order to proceed, we need to introduce some functional spaces.

We denote by $C_b(H)$ the Banach space of the bounded real valued and continuous function on H endowed with the usual supremum norm $\|\cdot\|_0$. We also denote by $C_{b,1}(H)$ the Banach space of all continuous functions $f : H \to \mathbb{R}$ such that

$$\|f\|_{0,1} := \|(1 + |\cdot|_2)^{-1} f\|_0 < \infty.$$

Now set

$$V(x) := |x|_6^8 |x|_4^2, \quad x \in L^6(0,1)$$

and denote by $C_{b,V}(L^6(0,1))$ the space of all continuous function $\varphi : L^6(0,1) \to \mathbb{R}$ such that the function

$$L^6(0,1) \to \mathbb{R}, \quad x \mapsto \frac{\varphi(x)}{1 + V(x)}$$

is bounded. The space $C_{b,V}(L^6(0,1))$, endowed with the norm

$$\|\varphi\|_{0,V} := \sup_{x \in L^6(0,1)} \frac{|\varphi(x)|}{1 + V(x)}$$

is a Banach space.

As easily seen, $C_b(H) \subset C_{b,1}(H) \subset C_{b,V}(L^6(0,1))$ with continuous embedding.

For a sequence $(\varphi_n)_{n\mathbb{N}} \subset C_{b,V}(L^6(0,1))$ and $\varphi \in C_{b,V}(L^6(0,1))$ we shall use the notation

$$\lim_{n \to \infty} \frac{\varphi_n}{1 + V} \overset{\pi}{=} \frac{\varphi}{1 + V}$$

to say that

$$\lim_{n \to \infty} \frac{\varphi_n(x)}{1 + V(x)} = \frac{\varphi(x)}{1 + V(x)}, \quad \forall x \in L^6(0, 1)$$

and

$$\sup_{n \in \mathbb{N}} \|\varphi_n\|_{0,V} < \infty.$$

When the sequence has more than an index, the meaning of the limit is the same given in Definition 2.1.1.

The reason which justifies the introduction of the above spaces is that we are not able to prove that the transition semigroup associated to the mild solution of (7.2) acts on uniformly continuous functions. However, thanks to the estimate given in Proposition 7.1.2 we can define a semigroup of transition operators in $C_{b,V}(L^6(0, 1))$ by the formula

$$P_t\varphi(x) = \mathbb{E}[\varphi(X(t, x))], \quad t \geq 0, \, \varphi \in C_{b,V}(L^6(0, 1)), \, x \in L^6(0, 1),$$
(7.6)

where $X(t, x)$ is solution of (7.5) (see Proposition 7.4.2). We define its infinitesimal generator by setting

$$
\begin{cases}
D(K, C_{b,V}(L^6(0, 1))) = \Big\{ \varphi \in C_{b,V}(L^6(0, 1)) : \exists g \in C_{b,V}(L^6(0, 1)), \\
\qquad\qquad \lim_{t \to 0^+} \dfrac{P_t\varphi(x) - \varphi(x)}{t} = g(x), \, x \in L^6(0, 1), \\
\qquad\qquad \sup_{t \in (0,1)} \left\| \dfrac{P_t\varphi - \varphi}{t} \right\|_{0,V} < \infty \Big\} \\
K\varphi(x) = \lim_{t \to 0^+} \dfrac{P_t\varphi(x) - \varphi(x)}{t}, \\
\qquad\qquad \varphi \in D(K, C_{b,V}(L^6(0, 1))), \, x \in L^6(0, 1).
\end{cases}
$$
(7.7)

We notice that $\mathcal{M}_V(L^6(0, 1))$ coincides with the space of all finite Borel measures $\mu \in \mathcal{M}(H)$ such that

$$\int_{L^6(0,1)} V(x)|\mu|_{TV}(dx) < \infty.$$

The first result of the chapter is the generalization of Theorems 2.2.3, 2.3.1. We omit the proof which is very similar to the proof of Theorem 5.1.2

Theorem 7.2.1. *Let $(P_t)_{t \geq 0}$ be the semigroup defined by (7.6) and let us consider its infinitesimal generator $(K, D(K, C_{b,V}(L^6(0, 1))))$ given by*

(7.7). Then, the formula

$$\langle \varphi, P_t^* F \rangle_{\mathcal{L}(\mathcal{C}_{b,V}(L^6(0,1)), (\mathcal{C}_{b,V}(L^6(0,1)))^*)}$$
$$= \langle P_t\varphi, F \rangle_{\mathcal{L}(\mathcal{C}_{b,V}(L^6(0,1)), (\mathcal{C}_{b,V}(L^6(0,1)))^*)}$$

defines a semigroup $(P_t^*)_{t\geq 0}$ of linear and continuous operators on $\mathcal{C}_{b,V}(L^6(0,1))$ which maps $\mathcal{M}_V(L^6(0,1))$ into $\mathcal{M}_V(L^6(0,1))$. Moreover, for any measure $\mu \in \mathcal{M}_V(L^6(0,1))$ there exists a unique family $\{\mu_t, t \geq 0\} \subset \mathcal{M}_V(L^6(0,1))$ such that

$$\int_0^T \left(\int_{L^6(0,1)} V(x)|\mu_t|_{TV}(dx) \right) dt < \infty, \quad \forall T > 0 \qquad (7.8)$$

and

$$\int_{L^6(0,1)} \varphi(x)\mu_t(dx) - \int_{L^6(0,1)} \varphi(x)\mu(dx)$$
$$= \int_0^t \left(\int_{L^6(0,1)} K\varphi(x)\mu_s(dx) \right) ds \quad (7.9)$$

for any $t \geq 0$, $\varphi \in D(K, \mathcal{C}_{b,V}(L^6(0,1)))$. Finally, the solution of (7.8), (7.9) is given by $P_t^*\mu, t \geq 0$.

We consider the *Kolmogorov* differential operator

$$K_0\varphi(x) = \frac{1}{2}\mathrm{Tr}[D^2\varphi(x)] + \langle x, AD\varphi(x)\rangle - \frac{1}{2}\langle D_\xi D\varphi(x), x^2\rangle, \quad (7.10)$$

$x \in L^6(0,1)$, $\varphi \in \mathcal{E}_A(H)$. The following result is the core of this chapter and it is proved in Section 7.8

Theorem 7.2.2. *The operator $(K, D(K, \mathcal{C}_{b,V}(L^6(0,1))))$ is an extension of K_0, and for any $\varphi \in \mathcal{E}_A(H)$ we have $\varphi \in D(K, \mathcal{C}_{b,V}(L^6(0,1)))$ and $K\varphi = K_0\varphi$. Finally, the set $\mathcal{E}_A(H)$ is a π-core for $(K, D(K, \mathcal{C}_{b,V}(L^6(0, 1))))$, that is for any $\varphi \in D(K, \mathcal{C}_{b,V}(L^6(0,1)))$ there exist $m \in \mathbb{N}$ and an m-indexed sequence $(\varphi_{n_1,...,n_m})_{n_1\in\mathbb{N},...,n_m\in\mathbb{N}} \subset \mathcal{E}_A(H)$ such that*

$$\lim_{n_1\to\infty} \cdots \lim_{n_m\to\infty} \frac{\varphi_{n_1,...,n_m}}{1+V} \overset{\pi}{=} \frac{\varphi}{1+V}$$

and

$$\lim_{n_1\to\infty} \cdots \lim_{n_m\to\infty} \frac{K_0\varphi_{n_1,...,n_m}}{1+V} \overset{\pi}{=} \frac{K\varphi}{1+V}.$$

The third main result of this chapter follows by the previous ones and by reasoning as for Theorems 5.1.4, 6.2.3.

Theorem 7.2.3. *For any $\mu \in \mathcal{M}_V(L^6(0,1))$ there exists an unique family of measures $\{\mu_t, \ t \geq 0\} \subset \mathcal{M}_V(L^6(0,1))$ fulfilling* (7.8) *and the measure equation*

$$\int_{L^6(0,1)} \varphi(x)\mu_t(dx) - \int_{L^6(0,1)} \varphi(x)\mu(dx)$$
$$= \int_0^t \left(\int_{L^6(0,1)} K_0\varphi(x)\mu_s(dx) \right) ds, \tag{7.11}$$

$t \geq 0$, $\varphi \in \mathcal{E}_A(H)$. *Moreover, the solution is given by* $P_t^* \mu$, $t \geq 0$.

7.3. Further estimates on the solution

In the case of the Burgers equation we are not able to prove uniform continuity of the solution with respect to the initial data as done in (5.3) in the case of Lipschitz nonlinearities or in (6.8) in the case of polynomial nonlinearities. We prove only uniform continuity with respect to the inital datum in a bounded neighbourhood. In order to proceed, set

$$\theta = \sup_{t\in[0,T]} |W_A(t)|_\infty, \quad T > 0.$$

Clearly θ is a random variable, and $\theta < \infty$ a.s.

We need the following estimates, proved in Lemma 3.1 of [20]

Lemma 7.3.1. *For any $p \in [2, \infty)$ there exists $c_p > 0$ such that if $Y(t,x)$ is a solution of* (7.5), *then*

$$|Y(t,x)|_p \leq c_p \left(\theta^3 + |x|_p \right) e^{1+2p\theta t}.$$

We have the following

Theorem 7.3.2. *For any $p \in [2, \infty)$ there exists a continuous function $c_p : (\mathbb{R}^+)^4 \to \mathbb{R}^+$ such that*

$$|Y(t,x) - Y(t,y)|_p \leq c_p(t, |x|_p, |y|_p, \theta)|x - y|_p, \quad x, y \in L^p(0,1).$$

Proof. Here we follow [20]. By (7.5) we have

$$Y(t,x) - Y(t,y) = e^{tA}(x-y)$$
$$+ \frac{1}{2} \int_0^t e^{(t-s)A} \frac{\partial}{\partial \xi} \big((Y(s,x) - Y(s,y))(Y(s,x) + Y(s,y) + 2W(s)) \big) ds.$$

then

$$|Y(t,x) - Y(t,y)|_p \le |x - y|_p$$
$$+ \frac{1}{2} \int_0^t \left| e^{(t-s)A} \frac{\partial}{\partial \xi} ((Y(s,x) - Y(s,y)) \right.$$
$$\left. (Y(s,x) + Y(s,y) + 2W(s))) \right|_p ds.$$
$$(7.12)$$

As well known, e^{tA}, $t \ge 0$ has smoothing properties. In particular, for any $s_1, s_2 \in \mathbb{R}$, $s_1 \le s_2, r \ge 1$, e^{tA} maps $W^{s_1,r}(0,1)$ into $W^{s_2,r}(0,1)$, for any $t > 0$. Moreover, there exists $C_1 > 0$, depending on s_1, s_2, r, such that

$$|e^{tA}z|_{W^{s_2,r}(0,1)} \le C_1 \left(1 + t^{\frac{s_1-s_2}{2}}\right) |z|_{W^{s_1,r}(0,1)}, \quad z \in W^{s_1,r}(0,1), \quad (7.13)$$

see Lemma 3, Part I in [43]. Using the Sobolev embedding theorem we have

$$\left| e^{(t-s)A} \frac{\partial}{\partial \xi} ((Y(s,x) - Y(s,y))(Y(s,x) + Y(s,y) + 2W(s))) \right|_p$$
$$\le C_1 \left| e^{(t-s)A} \frac{\partial}{\partial \xi} ((Y(s,x) - Y(s,y))(Y(s,x) + Y(s,y) + 2W(s))) \right|_{W^{\frac{1}{p}, \frac{p}{2}}(0,1)}$$

and, thanks to the above estimate with $s_1 = -1, s_2 = 1/p, r = p/2$

$$\left| e^{(t-s)A} \frac{\partial}{\partial \xi} ((Y(s,x) - Y(s,y))(Y(s,x) + Y(s,y) + 2W(s))) \right|_p$$
$$\le C_1 C_2 \left(1 + (t-s)^{-\frac{1}{2} - \frac{1}{2p}}\right) \times \left| \frac{\partial}{\partial \xi} ((Y(s,x) - Y(s,y))(Y(s,x) \right.$$
$$\left. + Y(s,y) + 2W(s))) \right|_{W^{-1, \frac{p}{2}}(0,1)}$$
$$\le C_1 C_2 \left(1 + (t-s)^{-\frac{1}{2} - \frac{1}{2p}}\right) \left| (Y(s,x) - Y(s,y))(Y(s,x) + Y(s,y) \right.$$
$$\left. + 2W(s)) \right|_{\frac{p}{2}}$$
$$\le C_1 C_2 \left(1 + (t-s)^{-\frac{1}{2} - \frac{1}{2p}}\right) |Y(s,x) - Y(s,y)|_p \left| Y(s,x) + Y(s,y) \right.$$
$$\left. + 2W(s) \right|_p$$
$$\le c_p C_1 C_2 \left(1 + (t-s)^{-\frac{1}{2} - \frac{1}{2p}}\right) |Y(s,x) - Y(s,y)|_p$$
$$\times \left((2\theta^3 + |x|_p + |y|_p) e^{1 + 2p\theta s} + 2\right).$$

Now the result follows by (7.12) and by Gronwall lemma (see, for instance, Lemma 7.1.1 in [32]). □

By recalling that $Y(t, x) = X(t, x) - W_A(t)$ it follows immediately the following result, which will be fundamental in the next section

Corollary 7.3.3. *For any* $p \in [2, \infty)$, $x \in L^p(0, 1)$, $T > 0$

$$\sup_{t \in [0,T]} |X(t, x + h) - X(t, x)|_p \to 0 \quad a.s., \ as \ |h|_p \to 0.$$

7.4. The transition semigroup in $\mathcal{C}_{b,V}(L^6(0, 1))$

This section is devoted in studying the semigroup $(P_t)_{t \geq 0}$ in the space $\mathcal{C}_{b,V}(L^6(0, 1))$.

Remark 7.4.1. By Corollary 7.3.3 we have that for any $\varphi \in \mathcal{C}_{b,V}(L^6(0,1))$

$$\lim_{\varepsilon \to 0} \sup_{|h|_6 < \varepsilon, \, t \in [0,T]} |P_t \varphi(x + h) - P_t \varphi(x)| = 0.$$

This is not sufficient to show that P_t, $t \geq 0$ maps uniformly continuous functions into uniformly continuous functions. However, as we shall see in (i) of the next proposition, this allows us to show that P_t maps the space $\mathcal{C}_{b,V}(L^6(0, 1))$ into itself.

Proposition 7.4.2. *Formula* (7.6) *defines a semigroup of operators* $(P_t)_{t \geq 0}$ *in* $\mathcal{C}_{b,V}(L^6(0, 1))$ *and there exist two constants* $c_0 \geq 1$, $\omega_0 \in \mathbb{R}$ *and a family of probability measures* $\{\pi_t(x, \cdot), \ t \geq 0, \ x \in L^6(0, 1)\} \subset \mathcal{M}_V(L^6(0, 1))$ *such that*

(i) $P_t \in \mathcal{L}(\mathcal{C}_{b,V}(L^6(0, 1)))$ *and* $\|P_t\|_{\mathcal{L}(\mathcal{C}_{b,V}(L^6(0,1)))} \leq c_0 e^{\omega_0 t}$;

(ii) $P_t \varphi(x) = \displaystyle\int_H \varphi(y) \pi_t(x, dy)$, *for any* $t \geq 0$, $\varphi \in \mathcal{C}_{b,V}(L^6(0, 1))$, $x \in L^6(0, 1)$;

(iii) *for any* $\varphi \in \mathcal{C}_{b,V}(L^6(0, 1))$, $x \in L^6(0, 1)$, *the function* $\mathbb{R}^+ \to \mathbb{R}$, $t \mapsto P_t \varphi(x)$ *is continuous.*

(iv) $P_t P_s = P_{t+s}$, *for any* $t, s \geq 0$ *and* $P_0 = I$;

(v) *for any* $\varphi \in \mathcal{C}_{b,V}(L^6(0,1))$ *and any sequence* $(\varphi_n)_{n \in \mathbb{N}} \subset \mathcal{C}_{b,V}(L^6(0,1))$ *such that*

$$\lim_{n \to \infty} \frac{\varphi_n}{1 + V} \stackrel{\pi}{=} \frac{\varphi}{1 + V}$$

we have, for any $t \geq 0$,

$$\lim_{n \to \infty} \frac{P_t \varphi_n}{1 + V} \stackrel{\pi}{=} \frac{P_t \varphi}{1 + V}.$$

Proof. (i). Take $\varphi \in C_{b,V}(L^6(0, 1))$, $t > 0$. We have to show that $P_t\varphi \in C_{b,V}(L^6(0, 1))$. By Proposition 7.1.2 it follows that

$$|P_t\varphi(x)| \leq \|\varphi\|_{0,V}(1 + \mathbb{E}[V(X(t, x))]) \leq c\|\varphi\|_{0,V}(1 + V(x)),$$

for some $c > 0$. Then, we have to show that the function $L^6(0, 1) \to \mathbb{R}$. $x \mapsto P_t\varphi(x)$ is continuous. Fix $x_0 \in L^6(0, 1)$. We have

$$|P_t\varphi(x_0 + h) - P_t\varphi(x_0)| \leq \mathbb{E}[\varphi(X(t, x_0 + h)) - \varphi(X(t, x_0))|].$$

By Corollary 7.3.3 we have that $|X(t, x_0 + h) - X(t, x_0)|_6 \to 0$ $\mathbb{P}$-a.s. as $|h|_6 \to 0$. Then, by the continuity of φ it follows $|\varphi(X(t, x_0 + h)) - \varphi(X(t, x_0))| \to 0$ $\mathbb{P}$-a.s. as $|h|_6 \to 0$. On the other hand, $\varphi(X(t, x_0+h))$ has bounded expectation, uniformly in any $L^6(0, 1)$-ball of center x_0. Then, it follows that $P_t\varphi(x_0 + h) \to P_t\varphi(x_0)$ has $|h|_6 \to 0$. (i) is proved. The other statements follows by arguing as for Proposition 5.1.6. □

Proposition 7.4.3. *Let $X(t, x)$ be the mild solution of problem (7.2) and let $(P_t)_{t \geq 0}$ be the associated transition semigroups in the space $C_{b,V}$ $(L^6(0, 1))$ defined by (7.6). Let also $(K, D(K, C_{b,V}(L^6(0, 1))))$ be the associated infinitesimal generators, defined by (7.7). Then*

(i) *for any $\varphi \in D(K, C_{b,V}(L^6(0, 1)))$, we have $P_t\varphi \in D(K, C_{b,V}$ $(L^6(0, 1)))$ and $K P_t\varphi = P_t K\varphi$, $t \geq 0$;*

(ii) *for any $\varphi \in D(K, C_{b,V}(L^6(0, 1)))$, $x \in H$, the map $[0, \infty) \to \mathbb{R}$, $t \mapsto P_t\varphi(x)$ is continuously differentiable and $(d/dt)P_t\varphi(x) = P_t K\varphi(x)$;*

(iii) *for any $\varphi \in C_{b,V}(L^6(0, 1))$, $t > 0$, the function*

$$H \to \mathbb{R}, \quad x \mapsto \int_0^t P_s\varphi(x)ds$$

belongs to $D(K, C_{b,V}(L^6(0, 1)))$, and it holds

$$K\left(\int_0^t P_s\varphi ds\right) = P_t\varphi - \varphi;$$

(iv) *for any $\lambda > \omega_0$, where ω_0 is as in Proposition 7.4.2, the linear operator $R(\lambda, K)$ on $C_{b,V}(L^6(0, 1))$ defined by*

$$R(\lambda, K)f(x) = \int_0^\infty e^{-\lambda t} P_t f(x)dt, \quad f \in C_{b,V}(L^6(0,1)), \; x \in L^6(0,1)$$

satisfies, for any $f \in C_{b,V}(L^6(0, 1))$

$$R(\lambda, K) \in \mathcal{L}(C_{b,V}(L^6(0,1))), \qquad \|R(\lambda, K)\|_{\mathcal{L}(C_{b,V}(L^6(0,1)))} \leq \frac{c_0}{\lambda - \omega_0}$$

$$R(\lambda, K)f \in D(K, \mathcal{C}_{b,V}(L^6(0, 1))), \quad (\lambda I - K)R(\lambda, K)f = f,$$

where c_0 is as in Proposition 7.4.2. We call $R(\lambda, K)$ the resolvent of K at λ.

Proof. (i), (ii) follows by 7.7, Proposition 7.4.2 and may be proved as for Theorem 2.2.4.

Let us show (iii). First, we have to check that $\int_0^t P_s f\, ds$ belongs to $\mathcal{C}_{b,V}(L^6(0, 1))$. By (i) of Proposition 7.4.2, for any $x \in L^6(0, 1)$ we have

$$\left| \int_0^t P_s \varphi(x) ds \right| \le \|\varphi\|_{0,V} c_0 \int_0^t e^{\omega_0 s} ds (1 + V(x)).$$

then,

$$\sup_{x \in L^6(0,1)} \frac{1}{1 + V(x)} \left| \int_0^t P_s \varphi(x) ds \right| < \infty.$$

Now let us fix $\varepsilon > 0$, $x_0 \in L^6(0, 1)$ and take $\delta > 0$ such that

$$\sup_{s \in [0,t]} \sup_{\substack{h \in L^6(0,1) \\ |h|_6 < \delta}} |P_s \varphi(x_0 + h) - P_s \varphi(x_0)| < \frac{\varepsilon}{t}.$$

The constant $\delta > 0$ exists thanks to Remark 7.4.1. Therefore, for any $h \in L^6(0, 1)$, $|h|_6 < \delta$ we have

$$\left| \int_0^t P_s \varphi(x_0 + h) ds - \int_0^t P_s \varphi(x_0) ds \right| \le \int_0^t |P_t \varphi(x_0 + h) - P_t \varphi(x_0)|\, ds$$

$$< \varepsilon.$$

By the arbitrariness of x_0, it follows $\int_0^t P_s \varphi ds \in \mathcal{C}_{b,V}(L^6(0, 1))$. The rest of the proof is essentially the same done for Theorem 2.2.4. $\quad\square$

7.5. Proof of Theorem 7.2.1

We point out that we have introduced the theory of stochastically continuous Markov semigroup in spaces of uniformly continuous functions only for convenience. All the result of chapter 2 remains true if we replace $\mathcal{C}_b(E)$ by $\mathcal{C}_b(E)$, the Banach space of all the continuous functions $f : E \to \mathbb{R}$, where E is a separable Banach space. Then, if we consider the semigroup $(P_t)_{t \ge 0}$ in (7.24) restricted to the space $\mathcal{C}_b(L^6(0, 1))$, by Remark 7.4.1 we have $P_t : \mathcal{C}_b(L^6(0, 1)) \to \mathcal{C}_b(L^6(0, 1))$ and consequently $(P_t)_{t \ge 0}$ is a stochastically continuous Markov semigroup in $\mathcal{C}_b(L^6(0, 1))$.

Now set[1]

$$
\begin{cases}
D(K, C_b(L^6(0, 1))) = \Big\{ \varphi \in D(K, C_b(L^6(0, 1))) : \exists g \in C_b(L^6(0, 1)), \\
\qquad\qquad \lim_{t \to 0^+} \dfrac{P_t \varphi(x) - \varphi(x)}{t} = g(x), \ x \in L^6(0, 1), \\
\qquad\qquad \sup_{t \in (0,1)} \left\| \dfrac{P_t \varphi - \varphi}{t} \right\|_0 < \infty \Big\} \\
K\varphi(x) = \lim_{t \to 0^+} \dfrac{P_t \varphi(x) - \varphi(x)}{t}, \\
\qquad \varphi \in D(K, C_b(L^6(0, 1))), \ x \in L^6(0, 1).
\end{cases}
$$

(7.14)

As pointed out above, all results of Chapter 2 remain true with C_b $(L^6(0, 1))$ replacing $C_b(E)$. In particular, Theorem 2.2.3 and Theorem can be extended to $(P_t)_{t \geq 0}$ and its infinitesimal generator $(K, D(K, C_b(L^6(0, 1))))$. For the reader's convenience, we summarize these results in the following theorem.

Theorem 7.5.1. *The family of linear maps* $P_t^* : (C_b(E))^* \to (C_b(E))^*$, $t \geq 0$, *defined by the formula*

$$
\langle \varphi, P_t^* F \rangle_{\mathcal{L}(C_b(L^6(0,1)), (C_b(L^6(0,1)))^*)} = \langle P_t \varphi, F \rangle_{\mathcal{L}(C_b(L^6(0,1)), (C_b(L^6(0,1)))^*)},
$$

where $t \geq 0$, $F \in (C_b(L^6(0, 1)))^*$, $\varphi \in C_b(L^6(0, 1))$, *is a semigroup of linear operators on* $(C_b(L^6(0, 1)))^*$ *which acts on* $\mathcal{M}(L^6(0, 1))$.

Moreover, for any $\mu \in \mathcal{M}(L^6(0, 1))$ *there exists a unique family of measures* $\{\mu_t, \ t \geq 0\} \subset \mathcal{M}(L^6(0, 1))$ *fulfilling*

$$
\int_0^T |\mu_t|_{TV}(L^6(0, 1)) dt < \infty, \quad T > 0;
$$

$$
\int_{C_b(L^6(0,1))} \varphi(x) \mu_t(dx) - \int_{C_b(L^6(0,1))} \varphi(x) \mu(dx)
$$

$$
= \int_0^t \left(\int_{C_b(L^6(0,1))} K\varphi(x) \mu_s(dx) \right) ds,
$$

for any $\varphi \in D(K, C_b(L^6(0, 1)))$, $t \geq 0$, *and the solution is given by* $P_t^* \mu$, $t \geq 0$.

Thanks to this theorem, by reasoning as in the proof of Theorem 6.2.1 we get the desired result.   □

---

[1] Here $\| \cdot \|_0$ denotes the supremum norm of $C_b(L^6(0, 1))$.

7.6. The OU semigroup in $C_{b,V}(L^6(0,1))$

Here we consider the transition semigroup in $C_{b,V}(L^6(0,1))$ associated to the mild solution of (7.3). It is well known (see, for instance, [14]) the following result

Proposition 7.6.1. *that for any $p, k \geq 1$, $T > 0$ there exists a constant $c_{p,k,T} > 0$ such that*

$$\mathbb{E}\left[\sup_{t\in[0,T]} |Z(t,x)|_p^k\right] \leq c_{p,k,T}\left(1 + |x|_p^k\right). \tag{7.15}$$

This easily implies that for any $T > 0$ there exists $c_T > 0$ such that

$$\mathbb{E}\left[\sup_{t\in[0,T]} V(Z(s,x))\right] \leq c_T\left(1 + V(x)\right). \tag{7.16}$$

Then, for any $t \geq 0$, we define the Ornstein-Uhlenbeck semigroup $(R_t)_{t\geq 0}$ in $C_{b,V}(L^6(0,1))$ by setting

$$R_t\varphi(x) = \mathbb{E}\left[\varphi(Z(t,x))\right], \quad t \geq 0, \; \varphi \in C_{b,V}(L^6(0,1)), \; x \in L^6(0,1), \tag{7.17}$$

where $Z(t,x)$ is the mild solution of (7.3). Clearly, (7.16) shows that R_t acts on $C_{b,V}(L^6(0,1))$. It is obvious that all the result of Proposition 7.4.2 holds also for the OU semigroup $(R_t)_{t\geq 0}$. We define the infinitesimal generator of $(R_t)_{t\geq 0}$ in $C_{b,V}(L^6(0,1))$ by setting

$$
\begin{cases}
D(L, C_{b,V}(L^6(0,1))) = \Big\{\varphi \in C_{b,V}(L^6(0,1)) : \exists g \in C_{b,V}(L^6(0,1)), \\
\qquad \lim_{t\to 0^+} \dfrac{R_t\varphi(x) - \varphi(x)}{t} = g(x), \forall x \in L^6(0,1), \\
\qquad \sup_{t\in(0,1)} \left\|\dfrac{R_t\varphi - \varphi}{t}\right\|_{0,V} < \infty \Big\} \\
L\varphi(x) = \lim_{t\to 0^+} \dfrac{R_t\varphi(x) - \varphi(x)}{t}, \\
\qquad \varphi \in D(L, C_{b,V}(L^6(0,1))), \; x \in L^6(0,1).
\end{cases}
\tag{7.18}
$$

Remark 7.6.2. Since all the results of Proposition 7.4.2 hold for the OU semigroup, it follows that all the results of Proposition 7.4.3 hold for the OU semigroup and its infinitesimal generator in $C_{b,V}(L^6(0,1))$.

We set

$$L_0\varphi(x) = \frac{1}{2}\mathrm{Tr}[D^2\varphi(x)] + \langle x, AD\varphi(x)\rangle, \quad \varphi \in \mathcal{E}_A(H), \; x \in L^6(0,1).$$

Proposition 7.6.3. *We have $\mathcal{E}_A(H) \subset D(L, \mathcal{C}_{b,V}(L^6(0,1)))$, and $L\varphi = L_0\varphi$, for any $\varphi \in \mathcal{E}_A(H)$.*

Proof. Since Hypothesis 1.2.1 holds for the operators A and $B = I$, the restriction of R_t to $\mathcal{C}_{b,1}(H)$ generates a semigroup of operators in $\mathcal{C}_{b,1}(H)$ (*cf.* Section 5.3.1 and Remark 5.1.5). We still denote the semigroup by $(R_t)_{t\geq 0}$. Now let $(L, D(L, \mathcal{C}_{b,1}(H)))$ be the infinitesimal generator of the OU semigroup in the space $\mathcal{C}_{b,1}(H)$, defined as in (3.5). Since $\mathcal{E}_A(H) \subset D(L, \mathcal{C}_{b,1}(H))$, to conclude the proof it is suffient to show

$$D(L, \mathcal{C}_{b,1}(H)) = \{\varphi \in D(L, \mathcal{C}_{b,V}(L^6(0,1))) \cap \mathcal{C}_{b,1}(H) : L\varphi \in \mathcal{C}_{b,1}(H)\}. \tag{7.19}$$

Indeed, if $\varphi \in D(L, \mathcal{C}_{b,V}(L^6(0,1))) \cap \mathcal{C}_{b,1}(H)$ and $L\varphi \in \mathcal{C}_{b,1}(H)$, in order to show $\varphi \in D(L, \mathcal{C}_{b,1}(H))$ it is sufficient to show

$$\sup_{t\in(0,1]} \left\| \frac{R_t\varphi - \varphi}{t} \right\|_{0,1} < \infty.$$

For any $x \in H$ we have

$$R_t\varphi(x) - \varphi(x) = \int_0^t R_t\varphi(x)ds.$$

Hence, since $L\varphi \in \mathcal{C}_{b,1}(H)$ and by the local boundedness of R_t we have

$$\sup_{t\in(0,1]} \left\| \frac{R_t\varphi - \varphi}{t} \right\|_{0,1} \leq \sup_{t\in(0,1]} \|R_t\|_{\mathcal{L}(\mathcal{C}_{b,1}(H))} \|L\varphi\|_{0,1} < \infty.$$

The other inclusion is obvious. This proves (7.19). By Proposition 5.3.2 it follows that $L\varphi = L_0\varphi$, $\forall \varphi \in \mathcal{E}_A(H)$. $\qquad\square$

Remark 7.6.4. We stress that in this chapter we work with the Ornstein-Uhlenbeck semigroup $(R_t)_{t\geq 0}$ in spaces of *continuous* functions. As we have pointed out in Remark 5.1.5, all the results of Chapter 5 remain true with $\mathcal{C}_b(H)$ replacing $\mathcal{C}_b(H)$ and $\mathcal{C}_{b,1}(H)$ replacing $\mathcal{C}_{b,1}(H)$.

7.7. The approximated problem

It is convenient to consider the usual Galerkin approximations of equation (7.2). For any $m \in \mathbb{N}$ we define

$$b_m(x) = P_m b(P_m x), \quad x \in H$$

where

$$P_m = \sum_{i=1}^{m} e_i \otimes e_i, \quad m \in \mathbb{N}.$$

We consider the approximating problem

$$\begin{cases} dX^m(t) = (AX^m(t) + b_m(X^m(t)))dt + dW(t), \\ X^m(0) = x. \end{cases} \tag{7.20}$$

By setting $Y^m(t, x) = X^m(t, x) - W_A(t)$, the corresponding mild form is

$$Y^m(t, x) = e^{tA}x + \frac{1}{2}\int_0^t e^{(t-s)A} P_m D_\xi \left(P_m(Y^m(s, x) + W_A(s))\right)^2 ds. \tag{7.21}$$

Since for any $m \in \mathbb{N}$ the identity

$$\langle b_m(x), x \rangle = 0, \quad x \in H$$

holds, all the estimates of Proposition 7.1.2, 7.7.2 are uniform on m and we have the following result.

Theorem 7.7.1. *For any $x \in L^p(0, 1)$, $p \in [2, \infty)$ there exists a unique mild solution $X^m \in L^p(0, 1)$ of equation (7.20). Moreover, for any $x_0 \in L^p(0, 1)$, $\delta > 0$ and $T > 0$*

$$\lim_{m \to \infty} \sup_{\substack{|x-x_0|_p < \delta \\ t \in [0, T]}} |X^m(t, x) - X(t, x)|_p = 0.$$

As in (4.3) we denote by P_t^m the transition semigroup

$$P_t^m \varphi(x) = \mathbb{E}[\varphi(X^m(t, x))], \quad t \ge 0, \; \varphi \in C_{b, V}(L^6(0, 1)), \; x \in L^6(0, 1). \tag{7.22}$$

By a standard argument, we find that for any $C_{b, V}(L^6(0, 1))$ we have

$$\lim_{m \to \infty} \frac{P_t^m \varphi}{1 + V} = \frac{P_t \varphi}{1 + V}, \quad t \ge 0.$$

For any $m \in \mathbb{N}$, we define the infinitesimal generator of the semigroup $(P_t^m)_{t \ge 0}$ by

$$\begin{cases} D(K_m, C_{b, V}(L^6(0, 1))) = \Big\{ \varphi \in C_{b, V}(L^6(0, 1)) : \exists g \in C_{b, V}(L^6(0, 1)), \\ \qquad\qquad \lim_{t \to 0^+} \dfrac{P_t^m \varphi(x) - \varphi(x)}{t} = g(x), x \in L^6(0, 1), \\ \qquad\qquad \sup_{t \in (0, 1)} \left\| \dfrac{P_t^m \varphi - \varphi}{t} \right\|_{0, V} < \infty \Big\} \\ K_m \varphi(x) = \lim_{t \to 0^+} \dfrac{P_t \varphi(x) - \varphi(x)}{t}, \\ \qquad\qquad \varphi \in D(K_m, C_{b, V}(L^6(0, 1))), \; x \in L^6(0, 1). \end{cases} \tag{7.23}$$

It is clear that all the results of Propositions 7.4.2, 7.4.3 hold for $(P_t^m)_{t \ge 0}$ and for its infinitesimal generator $(K_m, D(K_m, C_{b, V}(L^6(0, 1))))$.

7.7.1. The differential $DP_t^m \varphi$

In the previous chapters, we derived the properties of the differential $DP_t\varphi$ of the transition semigroup directly from the estimates on the differential $X_x(t, x)$ of the solution $X(t, x)$. This method cannot be applied here, by the lack of informations about $X_x(t, x)$. In [19], it is proposed to consider a Kolmogorov operator with an additional potential term

$$K_0'\varphi(x) = K_0\varphi(x) - c|x|_4^4\varphi(x), \quad \varphi \in \mathcal{E}_A(H)$$

and the corresponding semigroup given by the Feynman-Kac formula

$$S_t\varphi(x) = \mathbb{E}\left[e^{-c\int_0^t |X(s,x)|_4^4 ds} \varphi(X(t, x)) \right].$$

By using a generalization of the Bismut-Elworthy formula (see [25]) and some estimates on $X_x(t, x)$ the authors are able to get estimates on $DS(t)\varphi$. Then, by the formula

$$P_t\varphi = S_t\varphi + c \int_0^t S_{t-s} \left(|\cdot|_4^4\varphi \right) ds$$

they get estimates on $DP_t\varphi$.

This method it has been succesfully used to get solutions of the for the $3D$-Navier-Stokes equation (see [18, 27]). It has been also used to get smoothing properties of the differential $DP_t\varphi$, with application to control problems (see, for instance, [15, 16, 39])

The following result is proved in Proposition 3.6 of [19].

Proposition 7.7.2. *There exists $\omega_1 > 0$ such that for any $m \in \mathbb{N}$, $t > 0$ and $\varphi \in C_b^1(H)$ with $D\varphi \in C_b(H; H^1(0, 1))$ we have $DP_t^m\varphi(x) \in H^1(0, 1)$ and*

$$|DP_t^m\varphi(x)|_{H^1(0,1)} \le \left(\|D\varphi\|_{C_b(H;H^1(0,1))} + c\|\varphi\|_0 \right) (1 + |x|_6)^8 e^{\omega_1 t}.$$

The following two results are essential for the proof of Theorem 7.2.2.

Proposition 7.7.3. *Take $\lambda > \omega_0$, ω_1, where ω_0 is as in Proposition 7.4.2 and ω_1 is as in Proposition 7.7.2. Let $f \in \mathcal{E}_A(H)$ and, for $m \in \mathbb{N}$ consider the function*

$$L^6(0, 1) \to \mathbb{R}, \quad x \mapsto \varphi(x) = \int_0^\infty e^{-\lambda t} P_t^m f(x) dt.$$

Then

(i) φ *is continuous, bounded and Fréchet differentiable in any* $x \in L^6(0, 1)$ *with continuous differential* $D\varphi \in C(L^6(0, 1); H^1(0, 1))$. *Moreover, it holds*

$$|D\varphi(x)|_{H^1(0,1)} \le \frac{1}{\lambda - \omega_1} \left(\|Df\|_{C_b(H; H^1(0,1))} + c\|f\|_0 \right) (1 + |x|_6)^8;$$
(7.24)

(ii) φ *belongs to* $D(L, C_{b,V}(L^6(0, 1))) \cap D(K_m, C_{b,V}(L^6(0, 1)))$ *and*

$$K_m\varphi(x) = L\varphi(x) - \frac{1}{2} \langle D_\xi P_m D\varphi(x), (P_m x)^2 \rangle, \quad \forall x \in L^6(0, 1).$$
(7.25)

Proof. Notice that the mild solution of (7.20) is defined for any $x \in H$ (cf. [20]). Then, the transition semigroup P_t can be defined in $C_b(H)$. So, since $f \in C_b(H)$, it follows $\varphi \in C_b(H)$. By Proposition (7.7.2) we have

$$|D\varphi(x)|_{H^1(0,1)} \le \int_0^\infty e^{-\lambda t} |D P_t f(x)|_{H^1(0,1)} dt$$

$$\le \int_0^\infty e^{-(\lambda - \omega_1)t} dt \left(\|Df\|_{C_b(H; H^1(0,1))} + c\|f\|_0 \right) (1 + |x|_6)^8$$

and (7.24) follows. Still by (7.24) we get $D\varphi \in C(L^6(0, 1); H^1(0, 1))$. Indeed, for any $x, h \in L^6(0, 1)$,

$$|D\varphi(x + h) - D_\xi D\varphi(x)|_{H^1(0,1)}$$

$$\le \frac{1}{\lambda - \omega_1} \left(\|Df(\cdot + h) - Df(\cdot)\|_{C_b(H; H^1(0,1))} \right.$$

$$\left. + c\|f(\cdot + h) - f(\cdot)\|_0 \right) (1 + |x|_6)^8.$$

Since $f \in C_b(H)$, and $Df \in C_b(H; H^1(0, 1))$, by uniform continuity it follows $|D\varphi(x + h) - D\varphi(x)|_{H^1(0,1)} \to 0$ as $|h|_6 \to 0$. This concludes the proof of (i).

Let us prove (ii). Since the semigroup $(P_t^m)_{t \ge 0}$ satisfies the statements of Proposition 7.4.2, it follows that its infinitesimal generator K_m enjoys the statements of Proposition 7.4.3. In particular, we have $\varphi = R(\lambda, K_m)f$ and therefore $\varphi \in D(K_m, C_{b,V}(L^6(0, 1)))$. Then we have to show that $\varphi \in D(L, C_{h,V}(L^6(0, 1)))$. Now let $(R_t)_{t \ge 0}$ be the OU semigroup (7.17) and let $(L, D(L, C_{b,V}(L^6(0, 1))))$ be its infinitesimal generator in $C_{b,V}(L^6(0, 1))$. Fix $x \in L^6(0, 1)$, $T > 0$ and for $t \in [0, T]$ set $X^m(t) = X^m(t, x)$, $Z(t) = Z(t, x)$. By (7.21), (7.3) we have

$$X^m(t) = Z(t) + \frac{1}{2} \int_0^t e^{(t-s)A} P_m D_\xi (P_m X^m(s))^2 ds$$

and consequently

$$P_t^m \varphi(x) = \mathbb{E}\left[\varphi(X^m(t))\right]$$
$$= \mathbb{E}\left[\varphi\left(Z(t) + \frac{1}{2}\int_0^t e^{(t-s)A} P_m D_\xi (P_m X^m(s))^2 ds\right)\right].$$

Notice that since $f \in C_b^1(H)$, by (7.24) we get that the function $L^6(0,1) \to \mathbb{R}, x \mapsto D\varphi(x)$ is continuous. Then, by Taylor formula we have

$$R_t \varphi(x) - \varphi(x) = P_t^m \varphi(x) - \varphi(x)$$
$$+ \frac{1}{2}\mathbb{E}\left[\int_0^1 \left\langle D\varphi(\xi X^m(t) + (1-\xi)Z(t)),\right.\right. \tag{7.26}$$
$$\left.\left.\int_0^t e^{(t-s)A} P_m D_\xi (P_m X^m(s))^2 ds\right\rangle d\xi\right].$$

We claim that

$$\lim_{t\to 0^+} \frac{1}{t}\mathbb{E}\left[\int_0^1 \left\langle D\varphi(\xi X^m(t) + (1-\xi)Z(t)),\right.\right.$$
$$\left.\left.\int_0^t e^{(t-s)A} P_m D_\xi (P_m X^m(s))^2 ds\right\rangle d\xi\right]$$
$$= -\langle D_\xi P_m D\varphi(x), (P_m x)^2\rangle \tag{7.27}$$

holds. By Theorem 7.1.1, for any $T > 0$ we can write

$$X(t) = x + \theta_1(t)$$
$$Z(t) = x + \theta_2(t), \quad t \in [0,T]$$

where $\theta_1(t), \theta_2(t) : \Omega \to H, t \in [0,T]$ are random variables such that $\theta_1, \theta_2 \in C([0,T]; H)$ a.s. and $\theta_1(0) = \theta_2(0) = 0$. On the other hand, by Proposition 7.7.3 we can write

$$D\varphi(x+z) = D\varphi(x) + \eta(z), \quad z \in H$$

where $\eta \in C(H, H^1(0,1))$ and $\eta(0) = 0$. With these notations we have

$$D\varphi(\xi X^m(t) + (1-\xi)Z(t)) = D\varphi(x + \xi\theta_1(t) + (1-\xi)\theta_2(t))$$
$$= D\varphi(x) + \eta(\xi\theta_1(t) + (1-\xi)\theta_2(t)).$$

Then

$$\lim_{t\to 0^+} \sup_{\xi\in[0,1]} |D\varphi(\xi X^m(t) + (1-\xi)Z(t)) - D\varphi(x)|_{H^1(0,1)}$$
$$= \lim_{t\to 0^+} \sup_{\xi\in[0,1]} \left|\eta(\xi\theta_1(t) + (1-\xi)\theta_2(t))\right|_{H^1(0,1)} = 0. \tag{7.28}$$

For any $t > 0$ we have

$$\left| \frac{1}{t} \int_0^t e^{(t-s)A} P_m D_\xi (P_m X^m(s))^2 ds - P_m D_\xi ((P_m x)^2) \right|_{W^{-1,2}(0,1)}$$

$$\leq \frac{1}{t} \int_0^t \left| e^{(t-s)A} P_m D_\xi ((P_m X^m(s))^2 - (P_m x)^2) \right|_{W^{-1,2}(0,1)} ds \qquad (7.29)$$

$$+ \frac{1}{t} \int_0^t \left| e^{(t-s)A} P_m D_\xi (P_m x)^2 - P_m D_\xi (P_m x)^2 \right|_{W^{-1,2}(0,1)} ds.$$

The first term on the right-hand side is bounded by

$$\frac{1}{t} \int_0^t \left| P_m D_\xi ((P_m X^m(s))^2 - (P_m x)^2) \right|_{W^{-1,2}(0,1)} ds$$

$$\leq \frac{1}{t} \int_0^t \left| (P_m X^m(s))^2 - (P_m x)^2 \right|_2 ds$$

$$\leq \frac{1}{t} \int_0^t \left| X^m(s) - x \right|_2 \left| X^m(s) + x \right|_2 ds.$$

Since $X^m \in C([0, T]; H)$ $\mathbb{P}$-a.s., it follows

$$\lim_{t \to 0^+} \frac{1}{t} \int_0^t \left| e^{(t-s)A} P_m D_\xi ((P_m X^m(s))^2 - (P_m x)^2) \right|_{W^{-1,2}(0,1)} ds$$
$$= 0, \quad \mathbb{P}\text{-a.s.}$$

Since the semigroup e^{tA}, $t \geq 0$ can be estended to a strongly continuous semigroup in $W^{-1,2}(0, 1)$, for the last term of (7.29) it holds

$$\lim_{t \to 0^+} \frac{1}{t} \int_0^t \left| e^{(t-s)A} P_m D_\xi (P_m x)^2 - P_m D_\xi (P_m x)^2 \right|_{W^{-1,2}(0,1)} ds = 0.$$

Hence, by (7.29) we have

$$\lim_{t \to 0^+} \left| \frac{1}{t} \int_0^t e^{(t-s)A} P_m D_\xi (P_m X^m(s))^2 ds - P_m D_\xi (P_m x)^2 \right|_{W^{-1,2}(0,1)}$$
$$= 0, \quad \mathbb{P}\text{-a.s.}$$

This, together with (7.25) and an integration by parts, implies

$$\lim_{t \to 0^+} \frac{1}{t} \int_0^1 \left\langle D\varphi(\xi X^m(t) + (1-\xi)Z(t)), \int_0^t e^{(t-s)A} P_m D_\xi (P_m X^m(s))^2 ds \right\rangle d\xi$$
$$= \langle D\varphi(x), P_m D_\xi (P_m x)^2 \rangle = -\langle D_\xi P_m D\varphi(x), (P_m x)^2 \rangle, \quad \mathbb{P}\text{-a.s.}$$
$$(7.30)$$

In order to obtain (7.28), it is sufficient to show that the terms in the above limit are dominated by an integrable random variable. Indeed, for any $t \in (0, T]$ we have

$$
\frac{1}{t} \int_0^1 \left\langle D\varphi(\xi X^m(t) + (1 - \xi)Z(t)), \int_0^t e^{(t-s)A} P_m D_\xi (P_m X^m(s))^2 ds \right\rangle d\xi
$$

$$
\leq \frac{1}{t} \left| \int_0^1 D\varphi(\xi X^m(t) + (1 - \xi)Z(t)) d\xi \right|_{H^1(0,1)}
$$

$$
\times \left| \int_0^t e^{(t-s)A} P_m D_\xi (P_m X^m(s))^2 ds \right|_{W^{-1,2}(0,1)}
$$

$$
\leq \int_0^1 \left| D\varphi(\xi X^m(t) + (1 - \xi)Z(t)) \right|_{H^1(0,1)} d\xi
$$

$$
\times \frac{1}{t} \int_0^t \left| e^{(t-s)A} P_m D_\xi (P_m X^m(s))^2 \right|_{W^{-1,2}(0,1)} ds
$$

$$
\leq I_1(t) \times I_2(t).
$$

Set

$$
C_\varphi = \left(\frac{\|Df\|_{C_b(H;H^1(0,1))} + c\|f\|_0}{\lambda - \omega_1} \right).
$$

By (7.24) we have

$$
\int_0^1 |D\varphi(\xi X^m(t, x) + (1 - \xi)Z(t, x))|_{H^1(0,1)} d\xi
$$

$$
\leq C_\varphi \left(1 + |\xi X^m(t, x) + (1 - \xi)Z(t, x)|_6^8 \right)
$$

$$
\leq C_\varphi \int_0^1 \left(1 + \xi |X^m(t, x)|_6^8 + (1 - \xi)|Z(t, x)|_6^8 \right) d\xi
$$

$$
\leq C_\varphi \left(1 + \sup_{t \in [0,T]} |X^m(t, x)|_6^8 + \sup_{t \in [0,T]} |Z(t, x)|_6^8 \right).
$$

Here we have used the convexity of the function $z \to |z|_6^8$. For $I_2(t)$ we have

$$
I_2(t) \leq \frac{c}{t} \int_0^t \left| (P_m X^m(s))^2 \right|_2 ds
$$

$$
\leq \frac{c}{t} \int_0^t |X^m(s)|_4^2 ds \leq c \sup_{t \in [0,T]} |X^m(t)|_4^2.
$$

Then, for any $t \in (0, T]$ we have

$$\frac{1}{t}\left|\int_0^1 \left\langle D\varphi(\xi X^m(t)+(1-\xi)Z(t)), \int_0^t e^{(t-s)A} P_m D_\xi (P_m X^m(s))^2 ds\right\rangle d\xi\right|$$

$$\leq cC_\varphi \left(1 + \sup_{t\in[0,T]} |X^m(t,x)|_6^8 + \sup_{t\in[0,T]} |Z(t,x)|_6^8\right) \left(\sup_{t\in[0,T]} |X^m(t)|_4^2\right).$$

(7.31)

Notice that by Propositions 7.1.2, (7.6.1) the random variable

$$g(x) := cC_\varphi \left(1 + \sup_{t\in[0,T]} |X^m(t,x)|_6^8 + \sup_{t\in[0,T]} |Z(t,x)|_6^8\right)$$

$$\left(\sup_{t\in[0,T]} |X^m(t)|_4^2\right)$$

(7.32)

belongs to $L^1(\Omega, \mathbb{P})$ and

$$\mathbb{E}[g(x)] \leq C\left(1 + |x|_6^8|x|_4^2\right)$$

(7.33)

for some $C > 0$. Consequently, since for any $t \in (0, T]$

$$\frac{1}{t}\left|\int_0^1 \left\langle D\varphi(\xi X^m(t)+(1-\xi)Z(t)), \int_0^t e^{(t-s)A} P_m D_\xi (P_m X^m(s))^2 ds\right\rangle d\xi\right|$$

$$\leq g(x),$$

by the dominated convergence theorem and by (7.30) follows (7.27) as claimed.

By (7.26), (7.27) and by the fact that $\varphi \in D(K_m, \mathcal{C}_{b,V}(L^6(0,1)))$ we obtain

$$\lim_{t\to 0^+} \frac{R_t\varphi(x)-\varphi(x)}{t} = K_m\varphi + \frac{1}{2}\left\langle D_\xi P_m D\varphi(x), (P_m x)^2\right\rangle, \quad \forall x \in L^6(0,1).$$

Now, by (7.31), (7.32), (7.33) we have

$$\sup_{t\in(0,T]} \left|\frac{R_t\varphi(x)-\varphi(x)}{t}\right| \leq \sup_{t\in(0,T]} \left|\frac{P_t^m\varphi(x)-\varphi(x)}{t}\right| + \mathbb{E}[g(x)]$$

$$\leq c(1 + V(x))$$

since $\varphi \in D(K_m, \mathcal{C}_{b,V}(L^6(0,1)))$. This implies $\varphi \in D(L, \mathcal{C}_{b,V}(L^6(0,1)))$ and (7.25) follows.   □

Proposition 7.7.4. *Fix* $m \in \mathbb{N}$, $f \in \mathcal{E}_A(H)$ *and let* φ *be as in Proposition 7.7.3. Then, there exist* $k \in \mathbb{N}$ *and a* k-*indexed sequence* $(\varphi_{n_1,\ldots,n_k})_{n_1 \in \mathbb{N},\ldots,n_k \in \mathbb{N}} \subset \mathcal{E}_A(H)$ *such that*

$$\lim_{n_1 \to \infty} \cdots \lim_{n_k \to \infty} \frac{\varphi_{n_1,\ldots,n_k}}{1+V} \stackrel{\pi}{=} \frac{\varphi}{1+V} \tag{7.34}$$

$$\lim_{n_1 \to \infty} \cdots \lim_{n_k \to \infty} \frac{L_0\varphi_{n_1,\ldots,n_k}}{1+V} \stackrel{\pi}{=} \frac{L\varphi}{1+V} \tag{7.35}$$

and, for any $h \in H$

$$\lim_{n_1 \to \infty} \cdots \lim_{n_k \to \infty} \frac{\langle D_\xi D\varphi_{n_1,\ldots,n_k}, h \rangle}{\left(1+|\cdot|_6^8\right)} \stackrel{\pi}{=} \frac{\langle D_\xi D\varphi, h \rangle}{\left(1+|\cdot|_6^8\right)}. \tag{7.36}$$

Proof. Set

$$\psi_p(x) = \left(1 + p^{-1}|e^{\frac{1}{p}A}x|_6^8\right)^{-1} \varphi(e^{\frac{1}{p}A}x), \quad x \in H, \ p \in \mathbb{N}.$$

Clearly,

$$\lim_{p \to \infty} \frac{\psi_p}{1+|\cdot|_6^8} \stackrel{\pi}{=} \frac{\varphi}{1+|\cdot|_6^8}. \tag{7.37}$$

By the well known properties of the heat semigroup, $e^{\frac{1}{p}A}x \in L^6(0,1)$, for any $x \in H$. Then, since by Proposition 7.7.3 we have $\varphi \in C_b(L^6(0,1))$, it follows that $\psi_p : H \to \mathbb{R}$ is bounded. Moreover, an easy computation show that ψ_p is continuous. Then, $\psi_p \in C_b(H)$. A standard computation show

$$\langle D\psi_p(x), h \rangle = \frac{\langle D\varphi(e^{\frac{1}{p}A}x), e^{\frac{1}{p}A}h \rangle}{1 + p^{-1}|e^{\frac{1}{p}A}x|_6^8} - \frac{8\varphi(e^{\frac{1}{p}A}x)|e^{\frac{1}{p}A}x|_6^7\langle(e^{\frac{1}{p}A}x)^5, e^{\frac{1}{p}A}h \rangle}{p\left(1 + p^{-1}|e^{\frac{1}{p}A}x|_6^8\right)^2},$$

$x, h \in H$. Hence, by taking into account (7.24), there exists $c_f > 0$, depending on f, such that for any $x \in L^6(0,1)$ we have

$$|\langle D\psi_p(x), h \rangle|$$

$$\leq \frac{|D\varphi(e^{\frac{1}{p}A}x)|_2|e^{\frac{1}{p}A}h|_2}{1 + p^{-1}|e^{\frac{1}{p}A}x|_6^8} + \frac{8\|\varphi\|_0|e^{\frac{1}{p}A}x|_6^7|e^{\frac{1}{p}A}x|_6^3|e^{\frac{1}{p}A}h|_{L^{6/5}}}{p\left(1 + p^{-1}|e^{\frac{1}{p}A}x|_6^8\right)^2}$$

$$\leq \left(\frac{c_f(1 + |e^{\frac{1}{p}A}x|_6^8)}{\left(1 + p^{-1}|e^{\frac{1}{p}A}x|_6^8\right)(\lambda - \omega_1)} + \frac{2\|\varphi\|_0|e^{\frac{1}{p}A}x|_6^{10}}{p\left(1 + p^{-1}|e^{\frac{1}{p}A}x|_6^8\right)^2}\right)|e^{\frac{1}{p}A}h|_2$$

$$\leq \left(p\frac{c_f}{\lambda - \omega_1} + 2\|\varphi\|_0\right)|h|_2.$$

Then, ψ_p is Fréchet differentiable in any $x \in H$ and its differentiable is bounded. An easy but tedious computation shows that $D\psi_p : H \to \mathcal{L}(H)$ is continuous. Therefore[2], $\psi_p \in C_b^1(H)$. In addition, as easily checked, by (7.24) and by the above formulæ it follows

$$\lim_{p \to \infty} \frac{\langle D\psi_p, h \rangle}{1 + |\cdot|_6^8} \overset{\pi}{=} \frac{\langle D\varphi, h \rangle}{1 + |\cdot|_6^8}, \qquad \forall h \in H. \tag{7.38}$$

For any $n_2, n_3 \in \mathbb{N}$, consider the function

$$\psi_{n_2,n_3} : H \to \mathbb{R}, \qquad x \mapsto \psi_{n_2,n_3}(x) = n_2 \int_0^{\frac{1}{n_2}} R_t \psi_{n_3}(x) dt.$$

By Proposition 5.1.8, Remark 5.1.5 and by the above computation we find

$$\psi_{n_2,n_3} \in D(L, \mathcal{C}_{b,1}(H)) \cap C_b^1(H), \qquad n_2, n_3 \in \mathbb{N}.$$

Then, by Proposition 5.3.2 (cf. Remark 5.1.5 and Remark 7.6.4) there exists a sequence[3] $\{\psi_{n_2,n_3,n_4}\}_{n_4 \in \mathbb{N}} \subset \mathcal{E}_A(H)$ such that

$$\lim_{n_4 \to \infty} \psi_{n_2,n_3,n_4} \overset{\pi}{=} \psi_{n_2,n_3}, \qquad \lim_{n_4 \to \infty} L_0 \psi_{n_2,n_3,n_4} \overset{\pi}{=} L \psi_{n_2,n_3} \tag{7.39}$$

$$\lim_{n_4 \to \infty} \langle D\psi_{n_2,n_3,n_4}, h \rangle \overset{\pi}{=} \langle D\psi_{n_2,n_3}, h \rangle, \qquad \forall h \in H. \tag{7.40}$$

Now set

$$\varphi_{n_1} = R_{\frac{1}{n_1}} \varphi,$$

$$\varphi_{n_1,n_2} = n_2 \int_0^{\frac{1}{n_2}} R_{t + \frac{1}{n_1}} \varphi \, dt,$$

$$\varphi_{n_1,n_2,n_3} = n_2 \int_0^{\frac{1}{n_2}} R_{t + \frac{1}{n_1}} \psi_{n_3} \, dt,$$

$$\varphi_{n_1,n_2,n_3,n_4} = R_{\frac{1}{n_1}} \psi_{n_2,n_3,n_4}.$$

As easily checked, by the definition of $\varphi_{n_1,n_2,n_3,n_4}$ and by (7.37)

$$\lim_{n_1 \to \infty} \lim_{n_2 \to \infty} \lim_{n_3 \to \infty} \lim_{n_4 \to \infty} \frac{\varphi_{n_1,n_2,n_3,n_4}}{1 + |\cdot|_6^8} \overset{\pi}{=} \frac{\varphi}{1 + |\cdot|_6^8}.$$

which implies (7.34).

---

[2] $C_b^1(H)$ is the space of all $\varphi \in C_b(H)$ which are Fréchet differentiable with continuous and bounded differential $D\varphi : H \to \mathcal{L}(H)$.

[3] We assume that the sequence has one index.

Let us show (7.35). By (3.4) we have that $\varphi_{n_1,n_2,n_3,n_4} \in \mathcal{E}_A(H)$ and by Proposition 5.3.2 we have

$$L\varphi_{n_1,n_2,n_3,n_4} = L_0\varphi_{n_1,n_2,n_3,n_4}, \quad \forall n_1, n_2, n_3, n_4 \in \mathbb{N}.$$

Consequently, by (7.39) and by (i) of Proposition 5.1.8

$$\lim_{n_4 \to \infty} L_0\varphi_{n_1,n_2,n_3,n_4} = \lim_{n_4 \to \infty} R_{\frac{1}{n_1}} L\psi_{n_1,n_2,n_3}$$

$$\overset{\pi}{=} R_{\frac{1}{n_1}} L\psi_{n_1,n_2} = LR_{\frac{1}{n_1}} \psi_{n_1,n_2} = L\psi_{n_1,n_2,n_3}.$$

Still by Proposition 5.1.8 we have

$$LR_{\frac{1}{n_1}} \psi_{n_1,n_2} = n_2 \left(R_{\frac{1}{n_1}+\frac{1}{n_2}} \psi_{n_3} - R_{\frac{1}{n_1}} \psi_{n_3} \right).$$

Therefore

$$\lim_{n_2 \to \infty} \lim_{n_3 \to \infty} \frac{L\varphi_{n_1,n_2,n_3} \left(R_{\frac{1}{n_1}+\frac{1}{n_2}} \psi_{n_3} - R_{\frac{1}{n_1}} \psi_{n_3} \right)}{1+V}$$

$$\overset{\pi}{=} \lim_{n_2 \to \infty} \frac{\left(R_{\frac{1}{n_1}+\frac{1}{n_2}} \varphi - R_{\frac{1}{n_1}} \varphi \right)}{1+V}$$

$$\overset{\pi}{=} \frac{R_{\frac{1}{n_1}} L\varphi}{1+V}.$$

The last equality follows by (v) of Proposition 7.4.2 and by the fact that $\varphi \in D(L, \mathcal{C}_{b,V}(L^6(0,1)))$. Finally, since

$$\lim_{n_1 \to \infty} \frac{R_{\frac{1}{n_1}} L\varphi}{1+V} \overset{\pi}{=} \frac{L\varphi}{1+V},$$

(7.35) follows. Let us show (7.36). Notice that for any $n_1, n_2, n_3, n_4 \in \mathbb{N}$, $h \in H_0^1$ we have

$$\langle D\varphi_{n_1,n_2,n_3,n_4}(x), D_\xi h \rangle = R_{\frac{1}{n_1}} \left(\left\langle e^{\frac{1}{n_1}A} D\psi_{n_2,n_3,n_4}, D_\xi h \right\rangle \right)(x)$$

$$= -R_{\frac{1}{n_1}} \left(\left\langle D_\xi e^{\frac{1}{n_1}A} D\psi_{n_2,n_3,n_4}, h \right\rangle \right)(x).$$

By the elementary properties of the heat semigroup, for any $t > 0$ the linear operator $D_\xi e^{tA} : H_0^1 \to H$, $z \mapsto D_\xi e^{tA} z$ is bounded by $|D_\xi e^{tA} z|_2 \leq ct^{-1/2}|z|_2$, where $c > 0$ is indipendent of t. Then $D_\xi e^{\frac{1}{n_1}A} : H_0^1 \to H$ can be extended to a linear and bounded operator in H, which we still denote by $D_\xi e^{\frac{1}{n_1} A}$. Then by (7.40) we have

$$\lim_{n_4 \to \infty} \langle D_\xi D\varphi_{n_1,n_2,n_3,n_4}, h \rangle \stackrel{\pi}{=} \langle D_\xi D\varphi_{n_1,n_2,n_3}, h \rangle, \quad \forall h \in H.$$

By the same argument we find

$$\lim_{n_3 \to \infty} \langle D_\xi D\varphi_{n_1,n_2,n_3}, h \rangle \stackrel{\pi}{=} \langle D_\xi D\varphi_{n_1,n_2}, h \rangle, \quad \forall h \in H.$$

Notice now that by definition of φ_{n_1,n_2} we have

$$\langle D_\xi D\varphi_{n_1,n_2}(x), h \rangle = R_{\frac{1}{n_1}} \left(\left\langle D_\xi e^{\frac{1}{n_1} A} D\psi_{n_2}, h \right\rangle \right)(x), \quad x, h \in H.$$

Now, since $D_\xi e^{\frac{1}{n_1} A} : H \to H$ is linear and bounded, by (7.38) it follows

$$\lim_{n_2 \to \infty} \frac{\left\langle D_\xi e^{\frac{1}{n_1} A} D\psi_{n_2}, h \right\rangle}{1 + |\cdot|_6^8} \stackrel{\pi}{=} \frac{\left\langle D_\xi e^{\frac{1}{n_1} A} D\varphi, h \right\rangle}{1 + |\cdot|_6^8}.$$

Hence, by Proposition 7.4.2 (cf. Remark 7.6.2) we have

$$\lim_{n_2 \to \infty} \frac{\langle D_\xi D\varphi_{n_1,n_2}, h \rangle}{1 + |\cdot|_6^8} \stackrel{\pi}{=} \frac{\langle D_\xi D\varphi_{n_1}, h \rangle}{1 + |\cdot|_6^8}.$$

Finally, by Proposition 7.4.2 applied to the semigroup $(R_t)_{t \geq 0}$ we find

$$\lim_{n_1 \to \infty} \frac{\langle D_\xi D\varphi_{n_1}, h \rangle}{1 + |\cdot|_6^8} = \lim_{n_1 \to \infty} \frac{R_{\frac{1}{n_1}} \left(\langle D_\xi e^{\frac{1}{n_1} A} D\varphi, h \rangle \right)}{1 + |\cdot|_6^8}$$
$$\stackrel{\pi}{=} \frac{\langle D_\xi D\varphi, h \rangle}{1 + |\cdot|_6^8}.$$

This complete the proof. $\qquad \square$

7.8. Proof of Theorem 7.2.2

We split the proof into two lemmata.

Lemma 7.8.1. K is an extension of K_0 and $K\varphi = K_0\varphi$ for any $\varphi \in \mathcal{E}_A(H)$.

Proof. Take $h \in D(A)$. It is sufficient to show the claim for

$$\varphi(x) = e^{i\langle x,h\rangle}, \quad x \in L^6(0,1).$$

Let $(L, D(L, C_{b,1}(H)))$ be the infinitesimal generator in $C_{b,1}(H)$ of the Ornstein-Uhlenbeck semigroup associated to the mild solution of (7.3) and, for any $m \in \mathbb{N}$, let $(K_m, D(K_m, C_{b,V}(L^6(0,1))))$ be the infinitesimal generator of the semigroup $(P_t^m)_{t\geq 0}$ in $C_{b,V}(L^6(0,1))$, as defined in (7.22), (7.23). Since $\mathcal{E}_A(H) \subset D(L, C_{b,1}(H)) \cap C_{b,1}(H)$, by arguing as for Proposition 7.7.3 we find that for any $t \geq 0$, $x \in L^6(0,1)$ it holds

$$P_t^m \varphi(x) - R_t \varphi(x)$$
$$= \frac{i}{2}\mathbb{E}\left[\int_0^1 \varphi(\xi Z(t,x) + (1-\xi)X^m(t,x))d\xi \right.$$
$$\left\langle h, \int_0^t e^{(t-s)A} P_m D_\xi (P_m X^m(s,x))^2 ds\right\rangle\Bigg]$$
$$= \frac{i}{2}\mathbb{E}\left[\int_0^1 \varphi(\xi Z(t,x) + (1-\xi)X^m(t,x))d\xi \right.$$
$$\left.\int_0^t \left\langle h, e^{(t-s)A} P_m D_\xi (P_m X^m(s,x))^2\right\rangle ds\, d\xi\right]$$
$$= -\frac{i}{2}\mathbb{E}\left[\int_0^1 \varphi(\xi Z(t,x) + (1-\xi)X^m(t,x))d\xi \right.$$
$$\left.\int_0^t \left\langle D_\xi P_m e^{(t-s)A} h, (P_m X^m(s,x))^2\right\rangle ds\right],$$

since $D\varphi(x) = i\varphi(x)h$. Letting $m \to \infty$, by Theorem 7.7.1 we find

$$P_t \varphi(x) - \varphi(x) = R_t \varphi(x) - \varphi(x)$$
$$-\frac{i}{2}\mathbb{E}\left[\int_0^1 \varphi(\xi Z(t,x)+(1-\xi)X(t,x))d\xi \int_0^t \left\langle D_\xi e^{(t-s)A} h, (X(s,x))^2\right\rangle ds\right].$$

This implies, still by arguing as for Proposition 7.7.3,

$$\lim_{t\to 0^+} \frac{P_t \varphi(x) - \varphi(x)}{t} = L\varphi(x) - \frac{i}{2}\varphi(x)\langle D_\xi h, x^2\rangle$$
$$= L\varphi(x) - \frac{1}{2}\langle D_\xi D\varphi(x), x^2\rangle,$$

for any $x \in L^6(0,1)$. As easily seen, $|D_\xi e^{tA} h|_2 \leq \pi |D_\xi h|_2$, then

$$\left|\frac{P_t \varphi(x) - \varphi(x)}{t}\right| \leq \left|\frac{R_t \varphi(x) - \varphi(x)}{t}\right| + \frac{|D_\xi h|_2}{2t}\int_0^t \mathbb{E}[|X(s,x)|_4^2]ds$$

Now, since $\varphi \in D(L, C_{b,1}(H))$, the first term of right-hand side is bounded by

$$\left| \frac{R_t \varphi(x) - \varphi(x)}{t} \right| \leq c(1 + |x|_2),$$

where $c_{\varphi,T} > 0$ depends only by φ and T. By Proposition 7.1.2, the last term on the right-hand side is bounded by

$$\frac{|D_\xi h|_2}{2t} \int_0^t \mathbb{E}[|X(s,x)|_4^2]ds \leq \frac{|D_\xi h|_2}{2} \mathbb{E}[\sup_{t \in [0,T]} |X(t,x)|_4^2]ds$$

$$\leq \frac{c_T |D_\xi h|_2}{2}(1 + |x|_4^2),$$

where $c_T > 0$ depends only by T. Then,

$$\sup_{t \in (0,1)} \left\| \frac{P_t \varphi - \varphi}{t} \right\|_{0,V} < \infty.$$

This implies $\varphi \in D(K, C_{b,V}(L^6(0,1)))$ and $K\varphi = L\varphi - \frac{1}{2}\langle D_\xi D\varphi, (\cdot)^2 \rangle$. Consequently, the claim follows by Proposition 5.3.2. □

Lemma 7.8.2. $\mathcal{E}_A(H)$ *is a π-core for* $(K, D(K, C_{b,V}(L^6(0,1))))$, *that is for any* $\varphi \in D(K, C_{b,V}(L^6(0,1)))$ *there exist* $m \in \mathbb{N}$ *and an m-indexed sequence* $(\varphi_{n_1,\ldots,n_m})_{n_1 \in \mathbb{N}, \ldots, n_m \in \mathbb{N}} \subset \mathcal{E}_A(H)$ *such that*

$$\lim_{n_1 \to \infty} \cdots \lim_{n_m \to \infty} \frac{\varphi_{n_1,\ldots,n_m}}{1+V} \overset{\pi}{=} \frac{\varphi}{1+V} \tag{7.41}$$

and

$$\lim_{n_1 \to \infty} \cdots \lim_{n_m \to \infty} \frac{K_0 \varphi_{n_1,\ldots,n_m}}{1+V} \overset{\pi}{=} \frac{K\varphi}{1+V}. \tag{7.42}$$

Step 1. Take $\varphi \in D(K, C_{b,V}(L^6(0,1)))$ and fix $\lambda > \omega_0, \omega_1$, where ω_0 is as in Proposition 7.4.2 and ω_1 is as in Proposition 7.7.2. We set $\lambda \varphi - K\varphi = f$. By Proposition 7.4.3 we have $\varphi = R(\lambda, K)f$. Let us fix a sequence $(f_{n_1})_{n_1 \in \mathbb{N}} \subset \mathcal{E}_A(H)$ such that

$$\lim_{n_1 \to \infty} \frac{f_{n_1}}{1+V} \overset{\pi}{=} \frac{f}{1+V}.$$

We set $\varphi_{n_1} = R(\lambda, K)f_{n_1}$. By Proposition 7.4.2, 7.4.3 it follows

$$\lim_{n_1 \to \infty} \frac{\varphi_{n_1}}{1+V} \overset{\pi}{=} \frac{\varphi}{1+V}, \quad \lim_{n_1 \to \infty} \frac{K\varphi_{n_1}}{1+V} \overset{\pi}{=} \frac{K\varphi}{1+V}. \tag{7.43}$$

Step 2. Now let $(K_m, D(K_m, C_{b,V}(L^6(0,1))))$ be the infinitesimal generator of the semigroup $(P_t^m)_{t\geq 0}$ in the space $C_{b,V}(L^6(0,1))$, as defined in (7.23). We set

$$\varphi_{n_1,n_2} = \int_0^\infty e^{-\lambda t} P_t^{n_2} f_{n_1} dt.$$

By Proposition 7.4.3 we have $\varphi_{n_1,n_2} \in D(K_{n_2}, C_{b,V}(L^6(0,1)))$ and by a standard computation

$$\lim_{n_2\to\infty} \frac{\varphi_{n_1,n_2}}{1+V} \stackrel{\pi}{=} \frac{\varphi_{n_1}}{1+V}, \qquad \lim_{n_2\to\infty} \frac{K_{n_2}\varphi_{n_1,n_2}}{1+V} \stackrel{\pi}{=} \frac{K\varphi_{n_1}}{1+V}. \tag{7.44}$$

Notice that f_{n_1} satisfies the hypothesis of Proposition 7.7.3. Hence, $\varphi_{n_1,n_2} \in D(L, C_{b,V}(L^6(0,1)))$ and

$$K_{n_2}\varphi_{n_1,n_2} = L\varphi_{n_1,n_2} - \frac{1}{2}\langle D_\xi P_{n_2} D\varphi_{n_1,n_2}, (P_{n_2}\cdot)^2\rangle, \tag{7.45}$$

for any $n_1, n_2 \in \mathbb{N}$, $x \in L^6(0,1)$. In addition, by (7.24) it holds

$$\begin{aligned}
&\left|\langle D_\xi D\varphi_{n_1,n_2}(x), x^2\rangle - \langle D_\xi P_{n_2} D\varphi_{n_1,n_2}(x), (P_{n_2}x)^2\rangle\right| \\
&= \left|\langle D\varphi_{n_1,n_2}(x), D_\xi(x^2) - P_{n_2} D_\xi(P_{n_2}x)^2\rangle\right| \\
&\leq \left|D\varphi_{n_1,n_2}(x)\right|_{H^1(0,1)} \left|D_\xi(x^2) - P_{n_2} D_\xi(P_{n_2}x)^2\right|_{W^{-1,2}(0,1)} \\
&\leq \left(\frac{\|D\varphi_{n_1}\|_{C_b(H;H^1(0,1))} + c\|\varphi_{n_1}\|_0}{\lambda - \omega_1}\right)(1+|x|_6)^8 \\
&\quad \times \left|D_\xi(x^2) - P_{n_2} D_\xi(P_{n_2}x)^2\right|_{W^{-1,2}(0,1)}
\end{aligned}$$

for any $x \in L^6(0,1)$, where $W^{-1,2}(0,1)$ is the topological dual of H^1 endowed with the norm $|\cdot|_{W^{-1,2}(0,1)}$. Consequently,

$$\lim_{n_2\to\infty} \frac{\langle D_\xi D\varphi_{n_1,n_2}(x), x^2\rangle - \langle D_\xi P_{n_2} D\varphi_{n_1,n_2}(x), (P_{n_2}x)^2\rangle}{1+V} \stackrel{\pi}{=} 0. \tag{7.46}$$

Step 3. By Proposition 7.7.4 for any $n_1, n_2 \in \mathbb{N}$ there exists a sequence (we assume that it has one index) $\{\varphi_{n_1,n_2,n_3}\}_{n_3\in\mathbb{N}} \subset \mathcal{E}_A(H)$ such that

$$\lim_{n_3\to\infty} \frac{\varphi_{n_1,n_2,n_3}}{1+V} \stackrel{\pi}{=} \frac{\varphi_{n_1,n_2}}{1+V} \tag{7.47}$$

$$\lim_{n_3\to\infty} \frac{L_0\varphi_{n_1,n_2,n_3}}{1+V} \stackrel{\pi}{=} \frac{L\varphi_{n_1,n_2}}{1+V} \tag{7.48}$$

and

$$\lim_{n_3\to\infty} \frac{\langle D_\xi D\varphi_{n_1,n_2,n_3}, h\rangle}{1+|\cdot|_6^8} \stackrel{\pi}{=} \frac{\langle D_\xi D\varphi_{n_1,n_2}, h\rangle}{1+|\cdot|_6^8}, \qquad \forall h \in H.$$

Then it follows

$$\lim_{n_3 \to \infty} \frac{\left\langle D_\xi P_{n_2} D\varphi_{n_1,n_2,n_3}, (\cdot)^2 \right\rangle}{1+V} \stackrel{\pi}{=} \frac{\left\langle D_\xi P_{n_2} D\varphi_{n_1,n_2}, (\cdot)^2 \right\rangle}{1+V}. \tag{7.49}$$

Step 4. By construction, $(\varphi_{n_1,n_2,n_3})_{n_1,n_2,n_3} \subset \mathcal{E}_A(H)$. By (7.43), (7.44), (7.47)

$$\lim_{n_1 \to \infty} \lim_{n_2 \to \infty} \lim_{n_3 \to \infty} \frac{\varphi_{n_1,n_2,n_3}}{1+V} \stackrel{\pi}{=} \frac{\varphi}{1+V}.$$

Hence (7.41) follows. Let us show (7.42). By Lemma 7.8.1, for any $n_1, n_2, n_3 \in \mathbb{N}$, $x \in L^6(0, 1)$ we have

$$K\varphi_{n_1,n_2,n_3}(x) = K_0\varphi_{n_1,n_2,n_3}(x) = L_0\varphi_{n_1,n_2,n_3}(x) - \frac{1}{2}\left\langle D_\xi \varphi_{n_1,n_2,n_3}(x), x^2 \right\rangle.$$

By (7.48), (7.49),

$$\lim_{n_3 \to \infty} \frac{K_0\varphi_{n_1,n_2,n_3}}{1+V} \stackrel{\pi}{=} \frac{L\varphi_{n_1,n_2} - \frac{1}{2}\left\langle D_\xi D\varphi_{n_1,n_2}, (\cdot)^2 \right\rangle}{1+V}.$$

By (7.45) it holds

$$L\varphi_{n_1,n_2} - \frac{1}{2}\left\langle D_\xi D\varphi_{n_1,n_2}, (\cdot)^2 \right\rangle$$
$$= K_{n_2} D\varphi_{n_1,n_2} + \frac{1}{2}\left\langle D_\xi P_{n_2} D\varphi_{n_1,n_2}, (P_{n_2}\cdot)^2 \right\rangle - \frac{1}{2}\left\langle D_\xi D\varphi_{n_1,n_2}, (\cdot)^2 \right\rangle.$$

By (7.44), (7.46)

$$\lim_{n_3 \to \infty} \frac{L\varphi_{n_1,n_2,n_3} - \frac{1}{2}\left\langle D_\xi D\varphi_{n_1,n_2,n_3}, (\cdot)^2 \right\rangle}{1+V} \stackrel{\pi}{=} \frac{K\varphi_{n_1,n_2}}{1+V}.$$

Finally, by (7.43), (7.44) we have

$$\lim_{n_1 \to \infty} \lim_{n_2 \to \infty} \frac{K\psi_{n_1,n_2}}{1+V} \stackrel{\pi}{=} \frac{K\varphi}{1+V}.$$

7.9. Proof of Theorem 7.2.3

Take $\mu \in \mathcal{M}_V(L^6(0, 1))$.

Existence. By Theorem 7.2.2 we have $P_s^* \mu \in \mathcal{M}_V(L^6(0,1))$, for any $s \geq 0$. For any $\varphi \in D(K, \mathcal{C}_{b,V}(L^6(0,1)))$ we have

$$\lim_{h \to 0^+} \frac{1}{h} \left(\int_{L^6(0,1)} P_s \varphi(x) \mu(dx) - \int_{L^6(0,1)} P_s \varphi(x) \mu(dx) \right)$$

$$= \lim_{h \to 0^+} \frac{1}{h} \left(\int_{L^6(0,1)} K P_h \varphi(x) P_s^* \mu(dx) - \int_{L^6(0,1)} \varphi(x) P_s^* \mu(dx) \right)$$

$$= \lim_{h \to 0^+} \int_{L^6(0,1)} \frac{P_h \varphi(x) - \varphi(x)}{h} P_s^* \mu(dx)$$

$$= \int_{L^6(0,1)} K \varphi(x) P_s^* \mu(dx)$$

$$= \int_{L^6(0,1)} K_0 \varphi(x) P_s^* \mu(dx).$$

$$(7.50)$$

Here we have used (iv) of 7.7, Proposition 7.4.2 and Theorem 7.2.2. We stress that the limit above holds by the fact that $P_s^* \mu \in \mathcal{M}_V(L^6(0,1))$ and by 7.7. Still by Proposition 7.4.2 the function

$$\mathbb{R}^+ \to \mathbb{R}, \quad s \mapsto \int_{L^6(0,1)} K_0 \varphi(x) P_s^* \mu(dx)$$

is continuous. Then, by integrating (7.50) in $[0, t]$ we find that $P_t^* \mu$, $t \geq 0$ fulfils (7.11). By (i) of Proposition 7.4.2 it follows $|P_t^* \mu|_{TV} \leq c_0 e^{\omega_0 t} |\mu|_{TV}$. Hence $P_t^* \mu$, $t \geq 0$ fulfils (7.8).

Uniqueness. Assume that $\{\mu_t, \ t \geq 0\}$ fulfils (7.8), (7.11). Take $\varphi \in \mathcal{C}_{b,V}(L^6(0,1))$. By Theorem 7.2.2 there exist $m \in \mathbb{N}$ and an m-indexed sequence $(\varphi_{n_1,\dots,n_m})_{n_1,\dots,n_m \in \mathbb{N}} \subset \mathcal{E}_A(H)$ such that

$$\lim_{n_1 \to \infty} \cdots \lim_{n_m \to \infty} \frac{\varphi_{n_1,\dots,n_m}}{1+V} \stackrel{\pi}{=} \frac{\varphi}{1+V}$$

and

$$\lim_{n_1 \to \infty} \cdots \lim_{n_m \to \infty} \frac{K_0 \varphi_{n_1,\dots,n_m}}{1+V} \stackrel{\pi}{=} \frac{K\varphi}{1+V}.$$

Then, since $\{\mu_t, \ t \geq 0\} \subset \mathcal{M}_V(L^6(0,1))$, by the dominated convergence theorem we have

$$\lim_{n_1 \to \infty} \cdots \lim_{n_m \to \infty} \left(\int_{L^6(0,1)} \varphi_{n_1,\dots,n_m}(x) \mu_t(dx) - \int_{L^6(0,1)} \varphi_{n_1,\dots,n_m}(x) \mu(dx) \right)$$

$$= \int_{L^6(0,1)} \varphi(x) \mu_t(dx) - \int_{L^6(0,1)} \varphi(x) \mu(dx)$$

Similarly, for ant $s \in [0, t]$ we have

$$\lim_{n_1 \to \infty} \cdots \lim_{n_m \to \infty} \int_{L^6(0,1)} K_0 \varphi_{n_1,\dots,n_m}(x) \mu_s(dx)$$

$$= \int_{L^6(0,1)} K \varphi(x) \mu_s(dx).$$

Threfore, by (7.8) we can still apply the dominated convergence theorem to find

$$\lim_{n_1 \to \infty} \cdots \lim_{n_m \to \infty} \int_0^t \left(\int_{L^6(0,1)} K_0 \varphi_{n_1,\dots,n_m}(x) \mu_s(dx) \right) ds$$

$$= \int_0^t \left(\int_{L^6(0,1)} K \varphi(x) \mu_s(dx) \right) ds.$$

Then, $\{\mu_t, \ t \geq 0\}$ is solution of (7.8) and (7.9), for any $\varphi \in C_{b,V}(L^6(0,1))$. But by Theorem 7.2.1 such a solution is unique, thus μ_t must concides with $P_t^* \mu$, $\forall t \geq 0$. The proof is complete. $\qquad \square$

References

[1] S. AGMON, *Lectures on elliptic boundary value problems*, Prepared for publication by B. Frank Jones, Jr. with the assistance of George W. Batten, Jr. Van Nostrand Mathematical Studies, No. 2, D. Van Nostrand Co., Inc., Princeton, N.J.-Toronto-London, 1965.

[2] S. ALBEVERIO and M. RÖCKNER, *Stochastic differential equations in infinite dimensions: solutions via Dirichlet forms*, Probab. Theory Related Fields **89** (1991), no. 3, 347–386.

[3] V. BARBU, G. DA PRATO and A. DEBUSSCHE, *The Kolmogorov equation associated to the stochastic Navier-Stokes equations in 2D*, Infin. Dimens. Anal. Quantum Probab. Relat. Top. **7** (2004), no. 2, 163–182.

[4] V. I. BOGACHEV, G. DA PRATO, M. RÖCKNER and W. STANNAT, *Uniqueness of solutions to weak parabolic equations for measures*, Bull. Lond. Math. Soc. **39** (2007), no. 4, 631–640.

[5] V. I. BOGACHEV, G. DA PRATO and M. RÖCKNER, *Existence of solutions to weak parabolic equations for measures*, Proc. London Math. Soc. (3) **88** (2004), no. 3, 753–774.

[6] V. I. BOGACHEV AND MICHAEL RÖCKNER, *A generalization of Khasminskii's theorem on the existence of invariant measures for locally integrable drifts*, Teor. Veroyatnost. i Primenen. **45** (2000), no. 3, 417–436.

[7] V. I. BOGACHEV and M. RÖCKNER, *Elliptic equations for measures on infinite-dimensional spaces and applications*, Probab. Theory Related Fields **120** (2001), no. 4, 445–496.

[8] S. CERRAI, *A Hille-Yosida theorem for weakly continuous semigroups*, Semigroup Forum **49** (1994), no. 3, 349–367.

[9] S. CERRAI, *Second order PDE's in finite and infinite dimension*, Lecture Notes in Mathematics, vol. 1762, Springer-Verlag, Berlin, 2001, A probabilistic approach.

[10] A. CHOJNOWSKA-MICHALIK and B. GOLDYS, *Nonsymmetric Ornstein-Uhlenbeck semigroup as second quantized operator*, J. Math. Kyoto Univ. **36** (1996), no. 3, 481–498.

[11] A. CHOJNOWSKA-MICHALIK and B. GOLDYS, *Symmetric Ornstein-Uhlenbeck semigroups and their generators*, Probab. Theory Related Fields **124** (2002), no. 4, 459–486.

[12] G. DA PRATO, S. KWAPIEŃ and J. ZABCZYK, *Regularity of solutions of linear stochastic equations in Hilbert spaces*, Stochastics **23** (1987), no. 1, 1–23.

[13] G. DA PRATO, *Kolmogorov equations for stochastic PDEs*, Advanced Courses in Mathematics. CRM Barcelona, Birkhäuser Verlag, 2004.

[14] G. DA PRATO and A. DEBUSSCHE, *Differentiability of the transition semigroup of the stochastic Burgers equation, and application to the corresponding Hamilton-Jacobi equation*, Atti Accad. Naz. Lincei Cl. Sci. Fis. Mat. Natur. Rend. Lincei (9) Mat. Appl. **9** (1998), no. 4, 267–277 (1999).

[15] G. DA PRATO and A. DEBUSSCHE, *Dynamic programming for the stochastic Burgers equation*, Ann. Mat. Pura Appl. (4) **178** (2000), 143–174.

[16] G. DA PRATO and A. DEBUSSCHE, *Dynamic programming for the stochastic Navier-Stokes equations*, M2AN Math. Model. Numer. Anal. **34** (2000), no. 2, 459–475, Special issue for R. Temam's 60th birthday.

[17] G. DA PRATO and A. DEBUSSCHE, *Maximal dissipativity of the Dirichlet operator corresponding to the Burgers equation*, Stochastic processes, physics and geometry: new interplays, I (Leipzig, 1999), CMS Conf. Proc., vol. 28, Amer. Math. Soc., Providence, RI, 2000, 85–98.

[18] G. DA PRATO and A. DEBUSSCHE, *Ergodicity for the 3D stochastic Navier-Stokes equations*, J. Math. Pures Appl. (9) **82** (2003), no. 8, 877–947.

[19] G. DA PRATO and A. DEBUSSCHE, *m-dissipativity of Kolmogorov operators corresponding to Burgers equations with space-time white noise*, Potential Anal. **26** (2007), no. 1, 31–55.

[20] G. DA PRATO, A. DEBUSSCHE and R. TEMAM, *Stochastic Burgers' equation*, NoDEA Nonlinear Differential Equations Appl. **1** (1994), no. 4, 389–402.

[21] G. DA PRATO and L. TUBARO, *Some results about dissipativity of Kolmogorov operators*, Czechoslovak Math. J. **51(126)** (2001), no. 4, 685–699.

[22] G. DA PRATO and L. TUBARO, *Some results about dissipativity of Kolmogorov operators*, Czechoslovak Math. J. **51(126)** (2001), no. 4, 685–699.

[23] G. DA PRATO and J. ZABCZYK, *Stochastic equations in infinite dimensions*, Encyclopedia of Mathematics and its Applications, vol. 44, Cambridge University Press, 1992.

[24] G. DA PRATO and J. ZABCZYK, *Ergodicity for infinite-dimensional systems*, London Mathematical Society Lecture Note Series, vol. 229, Cambridge University Press, 1996.

[25] G. DA PRATO and J. ZABCZYK, *Differentiability of the Feynman-Kac semigroup and a control application*, Atti Accad. Naz. Lincei Cl. Sci. Fis. Mat. Natur. Rend. Lincei (9) Mat. Appl. **8** (1997), no. 3, 183–188.

[26] G. DA PRATO and J. ZABCZYK, *Second order partial differential equations in Hilbert spaces*, London Mathematical Society Lecture Note Series, vol. 293, Cambridge University Press, 2002.

[27] A. DEBUSSCHE and C. ODASSO, *Markov solutions for the 3D stochastic Navier-Stokes equations with state dependent noise*, J. Evol. Equ. **6** (2006), no. 2, 305–324.

[28] K.-J. ENGEL and R. NAGEL, *One-parameter semigroups for linear evolution equations*, Graduate Texts in Mathematics, vol. 194, Springer-Verlag, New York, 2000.

[29] S. N. ETHIER and T. G. KURTZ, *Markov processes, characterization and convergence*, Wiley Series in Probability and Mathematical Statistics: Probability and Mathematical Statistics, John Wiley & Sons Inc., 1986.

[30] E. FABES, M. FUKUSHIMA, L. GROSS, C. KENIG, M. RÖCKNER and D. W. STROOCK, *Dirichlet forms*, Lecture Notes in Mathematics, vol. 1563, Springer-Verlag, Berlin, 1993, Lectures given at the First C.I.M.E. Session held in Varenna, June 8–19, 1992, Edited by G. Dell'Antonio and U. Mosco.

[31] B. GOLDYS and M. KOCAN, *Diffusion semigroups in spaces of continuous functions with mixed topology*, J. Differential Equations **173** (2001), no. 1, 17–39.

[32] D. HENRY, *Geometric theory of semilinear parabolic equations*, Lecture Notes in Mathematics, vol. 840, Springer-Verlag, Berlin, 1981.

[33] T. W. KÖRNER, *Fourier analysis*, 2 ed., Cambridge University Press, 1989.

[34] J.-M. LASRY and P.-L. LIONS, *A remark on regularization in Hilbert spaces*, Israel J. Math. **55** (1986), no. 3, 257–266.

[35] Z. M. MA and M. RÖCKNER, *Introduction to the theory of (nonsymmetric) Dirichlet forms*, Universitext, Springer-Verlag, Berlin, 1992.

[36] L. MANCA, *On a class of stochastic semilinear PDEs*, Stoch. Anal. Appl. **24** (2006), no. 2, 399–426.

[37] L. MANCA, *Kolmogorov equations for measures*, To appear on *Journal of Evolution Equations*, 2007.

[38] L. MANCA, *Measure-valued equations for Kolmogorov operators with unbounded coefficients*, Preprint, 2007.

[39] L. MANCA, *On dynamic programming approach for the 3d-Navier-Stokes equations*, To be published on Appl. Math. Optim., 2007.

[40] E. PRIOLA, *On a class of Markov type semigroups in spaces of uniformly continuous and bounded functions*, Studia Math. **136** (1999), no. 3, 271–295.

[41] M. RÖCKNER and Z. SOBOL, *A new approach to Kolmogorov equations in infinite dimensions and applications to stochastic generalized Burgers equations*, C. R. Math. Acad. Sci. Paris **338** (2004), no. 12, 945–949.

[42] M. RÖCKNER and Z. SOBOL, *Kolmogorov equations in infinite dimensions: well-posedness and regularity of solutions, with applications to stochastic generalized Burgers equations*, Ann. Probab. **34** (2006), no. 2, 663–727.

[43] F. ROTHE, *Global solutions of reaction-diffusion systems*, Lecture Notes in Mathematics, vol. 1072, Springer-Verlag, Berlin, 1984.

[44] W. STANNAT, *(Nonsymmetric) Dirichlet operators on L^1: existence, uniqueness and associated Markov processes*, Ann. Scuola Norm. Sup. Pisa Cl. Sci. (4) **28** (1999), no. 1, 99–140.

[45] D. W. STROOCK and S. R. SRINIVASA VARADHAN, *Multidimensional diffusion processes*, Grundlehren der Mathematischen Wissenschaften [Fundamental Principles of Mathematical Sciences], vol. 233, Springer-Verlag, Berlin, 1979.

[46] K. YOSIDA, *Functional analysis*, Classics in Mathematics, Springer-Verlag, 1995, Reprint of the sixth (1980) edition.

[47] J. ZABCZYK, *Symmetric solutions of semilinear stochastic equations*, Stochastic partial differential equations and applications, II (Trento, 1988), Lecture Notes in Math., vol. 1390, Springer, Berlin, 1989, 237–256.

Index

THESES

This series gathers a selection of outstanding Ph.D. theses defended at the Scuola Normale Superiore since 1992.

Published volumes

1. F. COSTANTINO, *Shadows and Branched Shadows of 3 and 4-Manifolds*, 2005. ISBN 88-7642-154-8

2. S. FRANCAVIGLIA, *Hyperbolicity Equations for Cusped 3-Manifolds and Volume-Rigidity of Representations*, 2005. ISBN 88-7642-167-x

3. E. SINIBALDI, *Implicit Preconditioned Numerical Schemes for the Simulation of Three-Dimensional Barotropic Flows*, 2007.
   ISBN 978-88-7642-310-9

4. F. SANTAMBROGIO, *Variational Problems in Transport Theory with Mass Concentration*, 2007. ISBN 978-88-7642-312-3

5. M. R. BAKHTIARI, *Quantum Gases in Quasi-One-Dimensional Arrays*, 2007. ISBN 978-88-7642-319-2

6. T. SERVI, *On the First-Order Theory of Real Exponentiation*, 2008.
   ISBN 978-88-7642-325-3

7. D. VITTONE, *Submanifolds in Carnot Groups*, 2008.
   ISBN 978-88-7642-327-7

8. A. FIGALLI, *Optimal Transportation and Action-Minimizing Measures*, 2008. ISBN 978-88-7642-330-7

9. A. SARACCO, *Extension Problems in Complex and CR-Geometry*, 2008. ISBN 978-88-7642-338-3

10. L. MANCA, *Kolmogorov Operators in Spaces of Continuous Functions and Equations for Measures*, 2008. ISBN 978-88-7642-336-9

Volumes published earlier

H.Y. FUJITA, *Equations de Navier-Stokes stochastiques non homogènes et applications*, 1992.

G. GAMBERINI, *The minimal supersymmetric standard model and its phenomenological implications*, 1993. ISBN 978-88-7642-274-4

C. DE FABRITIIS, *Actions of Holomorphic Maps on Spaces of Holomorphic Functions*, 1994. ISBN 978-88-7642-275-1

C. PETRONIO, *Standard Spines and 3-Manifolds*, 1995.
ISBN 978-88-7642-256-0

I. DAMIANI, *Untwisted Affine Quantum Algebras: the Highest Coefficient of* det H_η *and the Center at Odd Roots of 1*, 1996.
ISBN 978-88-7642-285-0

M. MANETTI, *Degenerations of Algebraic Surfaces and Applications to Moduli Problems*, 1996. ISBN 978-88-7642-277-5

F. CEI, *Search for Neutrinos from Stellar Gravitational Collapse with the MACRO Experiment at Gran Sasso*, 1996. ISBN 978-88-7642-284-3

A. SHLAPUNOV, *Green's Integrals and Their Applications to Elliptic Systems*, 1996. ISBN 978-88-7642-270-6

R. TAURASO, *Periodic Points for Expanding Maps and for Their Extensions*, 1996. ISBN 978-88-7642-271-3

Y. BOZZI, *A study on the activity-dependent expression of neurotrophic factors in the rat visual system*, 1997. ISBN 978-88-7642-272-0

M.L. CHIOFALO, *Screening effects in bipolaron theory and high-temperature superconductivity*, 1997. ISBN 978-88-7642-279-9

D.M. CARLUCCI, *On Spin Glass Theory Beyond Mean Field*, 1998.
ISBN 978-88-7642-276-8

G. LENZI, *The MU-calculus and the Hierarchy Problem*, 1998.
ISBN 978-88-7642-283-6

R. SCOGNAMILLO, *Principal G-bundles and abelian varieties: the Hitchin system*, 1998. ISBN 978-88-7642-281-2

G. ASCOLI, *Biochemical and spectroscopic characterization of CP20, a protein involved in synaptic plasticity mechanism*, 1998.
ISBN 978-88-7642-273-7

F. PISTOLESI, *Evolution from BCS Superconductivity to Bose-Einstein Condensation and Infrared Behavior of the Bosonic Limit*, 1998.
ISBN 978-88-7642-282-9

L. PILO, *Chern-Simons Field Theory and Invariants of 3-Manifolds*,1999.
ISBN 978-88-7642-278-2

P. ASCHIERI, *On the Geometry of Inhomogeneous Quantum Groups*, 1999. ISBN 978-88-7642-261-4

S. CONTI, *Ground state properties and excitation spectrum of correlated electron systems*, 1999. ISBN 978-88-7642-269-0

G. GAIFFI, *De Concini-Procesi models of arrangements and symmetric group actions*, 1999. ISBN 978-88-7642-289-8

N. DONATO, *Search for neutrino oscillations in a long baseline experiment at the Chooz nuclear reactors*, 1999. ISBN 978-88-7642-288-1

R. CHIRIVÌ, *LS algebras and Schubert varieties*, 2003.
ISBN 978-88-7642-287-4

V. MAGNANI, *Elements of Geometric Measure Theory on Sub-Rieman-nian Groups*, 2003. ISBN 88-7642-152-1

F.M. ROSSI, *A Study on Nerve Growth Factor (NGF) Receptor Expres-sion in the Rat Visual Cortex: Possible Sites and Mechanisms of NGF Action in Cortical Plasticity*, 2004. ISBN 978-88-7642-280-5

G. PINTACUDA, *NMR and NIR-CD of Lanthanide Complexes*, 2004.
ISBN 88-7642-143-2

Fotocomposizione "CompoMat" Loc. Braccone, 02040 Configni (RI) Italy
Finito di stampare nel mese di dicembre 2008
dalla CSR srl, Via di Pietralata, 157, 00158 Roma